World-Building and
the Early Modern Imagination

World-Building and the Early Modern Imagination

Edited by

Allison B. Kavey

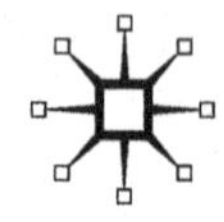

WORLD-BUILDING AND THE EARLY MODERN IMAGINATION

Softcover reprint of the hardcover 1st edition 2010 978-0-230-10588-1

First published in 2010 by
PALGRAVE MACMILLAN®
in the United States—a division of St. Martin's Press LLC,
175 Fifth Avenue, New York, NY 10010.

Where this book is distributed in the UK, Europe and the rest of the world, this is by Palgrave Macmillan, a division of Macmillan Publishers Limited, registered in England, company number 785998, of Houndmills, Basingstoke, Hampshire RG21 6XS.

Palgrave Macmillan is the global academic imprint of the above companies and has companies and representatives throughout the world.

ISBN 978-1-349-29014-7 ISBN 978-0-230-11313-8 (eBook)
DOI 10.1057/9780230113138

Library of Congress Cataloging-in-Publication Data

World-building and the early modern imagination / edited by Allison Kavey.
p. cm.

1. Europe—Intellectual life—16th century. 2. Europe—Intellectual life—17th century. 3. Philosophers—Europe—History—16th century. 4. Philosophers—Europe—History—17th century. 5. Philosophy, European—History—16th century. 6. Philosophy, European—History—17th century. 7. Imagination—History—16th century. 8. Imagination—History—17th century. I. Kavey, Allison, 1977–

CB401.W67 2010
940.2—dc22 2010010844

A catalogue record of the book is available from the British Library.

Design by Newgen Imaging Systems (P) Ltd., Chennai, India.

First edition: September 2010

10 9 8 7 6 5 4 3 2 1

For my parents, who encouraged me to imagine new worlds,
and for my friends, who join me to play in them.

Contents

Acknowledgments

Dane Daniel and I started talking about editing a book together on the imagination in early modern natural philosophy when we were both Neville Fellows at the Chemical Heritage Foundation in the summer of 2005. Despite jobs that took us to very different places and quickly took over our lives, we kept talking about the prevalence of imagination in early modern thought. Imagination shifted to world-building as I dug more deeply into Agrippa's *De Occulta Philosophia Libri Tres* and Dane grappled with Paracelsus's theology. We offered a series of panels on Renaissance world-building at a number of conferences in 2008 and 2009, and they provided the foundation for this volume. While Dane was not able to serve as coeditor of this book, it would not exist without his keen intelligence and innovative scholarship.

Much is also owed to the contributors to this volume, all of whom responded with enthusiasm to my requests that they think and write more about Renaissance world-building. Their essays reflect their exceptional talent and the dazzling array of sources that this period offers. They also demonstrate the ways in which modern disciplinary boundaries prove meaningless from an early modern perspective. In these pages, natural philosophy, drama, empire building, philosophy, and theology speak clearly to each other because they share so much. This book contains as many worlds as it discusses, and it reads so well because of it.

Chris Chappell, Samantha Hasey, and all the staff at Palgrave did an outstanding job of making this book a reality. Their consistent efficiency and enthusiasm for the project made it possible, and I thank them for their patience and generosity of spirit. The library staffs who cooperated with us on the images also deserve mention. The Folger Shakespeare Library, the Neville Collection at the Chemical Heritage Foundation, the New York Public Library, the Edinburgh University Library, and the Tate were exceptionally helpful.

INTRODUCTION

"Think you there was, or ever could be" a world such as this I dreamed

Allison B. Kavey

World-building is everywhere, or so it seems. From rebuilding Haiti after its devastating earthquake to the digital wonder-worlds on cinema screens, the twenty-first century imagination is obsessed with fixing old worlds and conjuring new ones. This is far from a new game. In the early modern period, people invented new worlds, created explanations for those they inhabited, and justified their relationships with other civilizations, nature, and God through their cosmogonical imaginations. Exploration and colonialism, both of which exploded during the sixteenth and seventeenth centuries, are evidently related to world-building. So, though perhaps less clearly, are the explanations scholars offered for nature's appearance and behavior, and the multitude of ties linking everything from God and the angels to sand fleas and pebbles. This volume explores the Renaissance preoccupation with world-building as it was practiced by natural philosophers, explorers, colonialists, and playwrights. The variety of perspectives illustrates the pervasiveness of world-building in this historical period, and it also illuminates the close ties among these intellectual spheres.

The Renaissance imagination transcended the disciplinary boundaries that modern scholars have used to study it. Scholars, explorers, playwrights, and theologians from this period exploited each others' ideas and intellectual frameworks as they built new worlds. Renaissance natural philosophers, for example, assembled worlds using ideas from the Old and New Testaments, ancient philosophers such as Aristotle and Plato, pseudo-ancient sources such as Hermes Trismegistus, medieval theologians, mathematicians, astrologers, and alchemists, and one another. They compiled these ideas in extraordinary new ways, producing new world systems that explained the way the natural world related to the Heavens and God, the forces operating to create natural change, and the ways in which those forces could be manipulated through natural or celestial magic to create desired effects. Theologians embarked on similar intellectual expeditions, exploiting the ideas offered by the Catholic Church from its infancy to the Renaissance, natural philosophers, natural historians such as Pliny, and even authors to conjure new worlds that illustrated God's role in the Creation and His ongoing relationship with humanity. The foot soldiers of the Church used these ideas about the origins of religion as fertile ground for conversion campaigns, creating imagined histories, uniting highly different civilizations into a coherent Christian nation. Explorers and colonialists built their own worlds, using what ancient authors and Renaissance navigators had to say to invent new maps of very old worlds, and employing a rich historical tapestry of ethnographic descriptions of the people they encountered to make sense of alien cultures. From Ireland to Africa to America, in fact, cosmogonical projects supported the military subjugation of individuals and nations deemed "new" or threatening by invading Europeans.

The dark side of world-building is inseparable from the wonder and imagination it reflects, and a coherent analysis of its place in Renaissance culture provides important insight into the intellectual schema of the period. The fact that it was so pervasive points to its utility. The sixteenth and seventeenth centuries were remarkable for the expansion that Europeans faced in nearly every aspect of their cultures, from the Reformation and Counter-Reformations, civil and international wars, exploration and colonialism, and the methodological exploitation of the natural world. Writers, scholars, soldiers, Jesuits, and rulers frequently encountered information that required radical reassessment of their worldviews. Sometimes, reassessment was not sufficient to assimilate newly acquired information with existing beliefs and interests, and as a result, new worlds were created. They offered their creators, from Drayton to Bacon, means of making sense out of dramatic changes in accepted truth without the requirement of abandoning strongly held convictions. In fact, one of the compelling things about the new worlds examined in this collection is their familiarity, which frequently sits comfortably beside startling new ideas. This alliance helped to perpetuate the proliferation of new worlds and it supported their entrance into popular culture, with playwrights, poets, and magicians printing, displaying, and performing their world-building enterprises throughout Europe.

Colonialism and the struggle to locate new worlds and create new social orders, for example, is intertwined with the literature from the period. Shakespeare's *Tempest*, which both borrowed from the English experience in Ireland and became an important subtext for future colonial experiments in the New World and Africa, is a perfect example of this intertwining. Guido Giglioni examines Shakespeare's magical island next to Bacon's

New Atlantis and Thomas More's Utopia to analyze the cultural utility of belief and its exercise in Renaissance thought. The extension of belief into new worlds, for these authors, allowed the reexamination of potentially damaging truths in imaginary spaces. The fact that these new worlds were consumed with such tremendous enthusiasm by the public and intellectuals alike reflects their powerful potential for exorcising cultural anxieties. Mark Waddell discusses another vastly public imagined world in his examination of Athanasius Kircher's museum. The incredible collection in this library of natural history, natural philosophy, and ethnography reflected its creator's vast enthusiasm for God's work, and as Waddell argues, systematized for the viewer a belief in God's continuing role in the world and His preeminence in the study of nature. Al Coppola, in his chapter on Thomas Burnet's *Sacred Theory of Earth*, also discusses the critical importance of the imagination in assimilating the apparently conflicting philosophies offered by Christianity and natural philosophy. While Burnet's book was rapidly rejected on the grounds of intellectual weakness and heresy, Coppola contends that it reflects the early modern conviction that prevailing belief systems needed reconciliation, and that imagined new worlds, studded with acknowledged truths, provided ripe ground for it.

Natural philosophers throughout Europe embarked on radical reassessments of their worlds to assimilate new information about nature with acknowledged truths and epistemological systems. This frequently resulted in accusations of heresy. The contentious physician, natural philosopher, and theologian Theophrastus Bombastus von Hohenheim Paracelsus, for example, proposed an entirely new version of mortal and immortal flesh, the Trinity, and the role of God in nature. Dane Daniel takes up Paracelsus's long overlooked theological writings, proposing that they shed critical light on his ideas about magic and the forces governing change in the natural world and concluding that the infamous German was as heretical as he was deemed by many of his opponents but for very different reasons. Allison Kavey takes up the writings of another magician and heretic, Heinrich Cornelius Agrippa von Nettesheim, whose opus magnus—*De Occulta Philosophia Libri Tres*—contained a new world that had a significant impact on magical and natural philosophical thought from the sixteenth century forward. Through close textual analysis, she concludes that Agrippa's text provided the script for a revised Creation, in which God intended, by providing the capacity for passion, knowledge, and imagination to humanity and littering nature with elaborate systems of sympathy and occult virtues, to lend magicians the chance to create their own worlds. Sheila Rabin examines another radical reimagination of the Creation, this time offered in the astrological and astronomical writings of Johannes Kepler. She concludes that the centrality of geometry to his intellectual framework provides the key to understanding his new version of the Creation and his poorly apprehended ideas about the practice of astrology.

Belief is also a critical theme for James De Lorenzi and Matteo Salvadore. Both resurrect the arguments employed by European Jesuits operating under the auspice of the Catholic Church and Ethiopian leaders to make sense of the cross-currents of religious belief that originally united and then violently divided these two groups. An examination of the variety of texts written both by European and Ethiopians during this period of contact and conflict reflects the contest between their cultural beliefs, and the ongoing process of cosmogony and world revision that occupied both parties. Belief was not, however, the only force motivating religious and political colonization. As Patrick Tuite and Vincent Carey contend, historical ideas of cultural and religious superiority mapped onto the English military, and social structure supported the violent subjugation of the Irish starting in the sixteenth century. Both authors use painfully clear visual and textual evidence from over two centuries to illustrate the ways in which English authors employed existing cultural tropes about the Irish to support their consistent reimagination as bestial, savage, pagan, and deserving of their fate in the hands of the English military.

From Kircher's graceful version of God's kingdom to the savage killing fields of Ireland, Catholic conversion in Ethiopia, magic and its relationship to God reconceived by Paracelsus and Agrippa, the Creation and the relationship between Heaven and Earth reimagined by Kepler and Burnet, and the utopias of Shakespeare, More, and Bacon, the Renaissance abounded with new worlds. This volume examines the plethora of cosmogonies that emerged during the sixteenth and seventeenth century. The authors ask what motivated this variety of institutions and individuals to engage in world-building, its cultural utility, and the receptions these new worlds received. Close textual and visual analysis provides the foundation for these chapters, and the array of sources illustrates the rich tapestry of ideas, anxieties, and enthusiasms that served as the basis for world-building. Only through investigating imagined worlds as closely as scholars have examined "real" Renaissance landscapes can we hope to understand the intellectual and cultural reassessments that characterized it, and the critical importance of imagination and belief in its intellectual landscape.

CHAPTER 1

Paracelsus on the "New Creation" and Demonic Magic: Misunderstandings, Oversights, and False Accusations in His Early Reception

Dane T. Daniel

Introduction

In his seventeenth-century *History of Magick* Gabriel Naudé delivered a mixed verdict regarding Paracelsus' efforts. He conceded that Theophrastus Bombast von Hohenheim, commonly called Paracelsus (1493/1494–1541), contributed positively to the course of medicine and science. Next, he denied any practical value to Paracelsus' magic—mostly he was properly arguing that Paracelsus, who wrote at length about theories and types of magic, should not be charged in any negative way for engaging in a nefarious practical magic. And then, as have historians for centuries, Naudé touched just briefly on Paracelsus' theology. Pointing to Paracelsus' braggadoccio, he labeled the Swiss-German iconoclast an "Arch-heretick," noting that "[Paracelsus threatened to bring] both the *Pope* and Luther... to his Maxims when he should think fit to do it."[1] This judgment was based not on his familiarity with Paracelsus' scriptural interpretations but rather on his lack of access to Paracelsus' theological writings and, perhaps even more so, the medical practitioner's reputation. The latter was influenced in part by the inaccurate claims of Heidelberg medical professor Thomas Erastus (1524–1583) that Paracelsus was an Arian (i.e., one who denies that Christ is of one substance with the Father and one who considers Jesus to be a creation by the Father) as well as a consorter with demons.[2]

Naudé's impression, influenced both by Paracelsian tracts and the criticisms by Paracelsus' detractors, was exemplary of the early modern understanding of Paracelsus. Paracelsus was noted for his iconoclastic contributions to medicine and natural philosophy as well as his dabbling in theoretical magic, the latter of which was well known in part because of its prodigious treatment in Paracelsus' opus magnum, the *Astronomia Magna* (1537/1538). In fact, Paracelsus married the fields of medicine, science, and magic with his discussion of such topics as medicinal virtues (e.g., *arcana*), sidereal powers in nature (including their role in instructing human beings via the "light of nature"), and inner alchemists that direct physiological processes (including digestion). The reputation that Paracelsus flirted with the demonic was not a strong one in that it was based, as I discuss, on a few opinions and could not be verified in the primary sources—the accusation of demonic dealings simply could not gain enough traction to damage him too much. And Paracelsus' *theologica*—found almost exclusively in manuscript rather than printed form—was very much misunderstood.

Erastus, for his part, had a very dubious plan for halting the spread of Paracelsian ideas; he attacked the Swiss-German medical practitioner for the wrong reasons. In addition to his ridiculous accusation that Paracelsus was a practitioner of the black arts, Erastus—on the basis of his reading of the *Philosophia ad Athenienses*, a text of dubious authenticity—chose to emphasize a mistaken interpretation of Paracelsus' Christology.[3] With even a little knowledge of Paracelsus' vast explicitly religious oeuvre, Erastus would not have overlooked the most contentious aspects of Paracelsus' theology. Drawing from pertinent passages within Paracelsus' authentic writings, I highlight Paracelsus' heretical theology, especially his concept of the cosmogony of God the Son, also called the "new creation," *id est* the creation of "immortal matter." (I do not treat Paracelsus' exegesis of the Genesis chapters on the creation of the natural world and "mortal matter" by God the Father.)[4] As I illustrate via an analysis of Paracelsus' Eucharistic tracts, the Paracelsians were lucky that Erastus and other detractors did not have access to Paracelsus' concept of the mortal flesh versus immortal flesh dichotomy, for it clearly falls within Gnostic and/or Docetic heresy.[5] I also explore Paracelsus' concept of magic, drawing attention to his ideas regarding "natural magic" in Book I of the *Astronomia Magna* and demonic magic in Book IV, wherein one finds a clear and exemplary indication of his opposition to dark powers.

An evaluation of Erastus' two questionable claims will help clarify Paracelsus' theology and approach to magic, and it promises to tell us much about Paracelsus' reception. I argue that despite the obvious heretical aspects of Paracelsus' theology, Paracelsus' followers seemed to overlook his controversial anthropological (referring to the nature of the human being) and Christological ideas, and in fact assumed him to be an orthodox Christian teacher. This misconception is evident in the popular seventeenth- and eighteenth-century work attributed to Paracelsus: *Kleine Hand und Denckbibel*.[6] There even seems to have been a cover-up of his heresy, an observation that Carlos Gilly has recently implied: "A number of Paracelsians…decided to dodge the specific theological issues."[7] The charge of demonic magic, on the other hand, seemingly did little to halt the momentum of Paracelsianism. It is more likely that Paracelsian tracts on magic helped to recruit followers to the cause of chymical medicine and Paracelsian philosophy; Paracelsus' discussion of magic in the *Astronomia Magna*, for example, was printed numerous times.[8]

The New Creation and Its Reception

Since a few decades after his death in 1541, some Paracelsian enthusiasts and several scholars have actually explored in some depth Paracelsus' idiosyncratic theology. Examples are the scribes who copied Paracelsus' numerous tracts on such topics as Mary and the Lord's Supper, and a few editors—such as Michael Toxites—who occasionally came across Paracelsus' eccentric biblical exegesis when working with such philosophical texts as the *Astronomia Magna*.[9] Other examples are certain followers of Valentin Weigel, who included Paracelsus' Eucharistic tracts in their diverse collections in the late 1610s, and then nineteenth- and twentieth-century scholars such as H.U. Preu and Kurt Goldammer who realized—unlike many who grapple with Paracelsus—that Paracelsus' thought as a whole cannot be well characterized or understood without a familiarity with his concept of the "two types of flesh."[10]

And yet, neither the followers nor detractors of Paracelsus drew any significant attention to his two-flesh concept, in which both Christ's body and the resurrection bodies of the saints consist not of earthly corporeal flesh, but rather a new type of spiritual, subtle flesh. Erastus' insubstantial argument, as noted above, taught that Paracelsus egregiously failed to make Christ coeval with the Father. Charles D. Gunnoe, Jr. has shown that Konrad Gessner (1516–1565) was among those who influenced Erastus' opinion. In the early 1560s Gessner produced a number of scathing accusations and warnings concerning the sect of Paracelsians; he also stressed their lack of both morality and education. Paracelsus, Gessner exclaimed, not only associated himself with demons, but also denied Christ's divine nature: the Paracelsians were Antitrinitarian Arians.[11] Gunnoe adds that Erastus, in his attack on Paracelsian medicine, honed in on the "Arianism" in the *Philosophia ad Athenienses*. Gunnoe explains:

> Seen in the light of Erastus' interpretation of Genesis, Paracelsus denied both the initial creation ex nihilo as well as God's role in the secondary acts of creation in giving composite materials their set forms. The Paracelsian notion that angelic or demonic forces were at work in this separation troubled Erastus, and he surmised that Paracelsus had really believed that Christ was one of these minor deities but Paracelsus did not have the courage to say it. In this connection, he accused Paracelsus of Arianism in placing the son in a subordinate position to the father.[12]

It must be noted, then, that Erastus drew much of his opinion from what is possibly a spurious text. Some scholars today do not consider the *Philosophy to the Athenians* to be an authentic text, but during the Early Modern Period this was not the case.[13] Again, access to virtually any of Paracelsus' authentic *theologica* would have dispelled the charge of Arianism. Indeed, in stark contrast to Gessner and Erastus' argument that Paracelsus was an Arian, Paracelsus emphasized Christ's divinity at the expense of His humanity.

The oversight of Paracelsus' concept of the new creation is puzzling, for Paracelsus' approach to Christ's flesh and the Christian immortal body are fundamental to his understanding of the nature of humans and the universe. He was obsessed with his bipartite approach to the flesh. Dozens of his tracts featured his theory that Christians possess not only a corporeal body, which is destined for irreversible destruction alongside all matter in the final conflagration, but also a resurrection body of perfect subtle flesh.[14] He taught in such tracts as *De genealogia Christi* that God the Father had created the mortal physical realm, but that God the Son had created the immortal realm of perfect subtle corporeality. The two types of flesh exist side by side in this world, but only Christ's creation will last. Experts of Paracelsus' *theologica*, such as Kurt Goldammer and Hartmut Rudolph, often refer to the two-flesh concept as Paracelsus' "Eucharistic thought" because he taught that the earth of the Father's creation provides the physical body with sustenance whereas the earth (or "limbus") of the Son's creation—which can be found in the bread and wine of the Lord's Supper—sustains the resurrection body.[15] Paracelsus writes:

> We are all from the earth, and the earth is that from which we are made. Now, does not the bread also come from the earth, and does not the wine also from of the earth? Yes. These things come from the same place as the human. It is not that the human is to eat the material earth in its substance, rather, the human eats the food (*speis*) that comes from it, and we imbibe the drink that comes from it. . . . Now [with regard to the second birth], as we are born from God—that is, from the body of Christ, from his bones and flesh—thus is he also the field (a*cker*) which gives the fruit of its body; [this fruit] that comes forth is food (*die nahrung*). Now, it follows that we do not eat Christ in the person; in contrast, we consume the food and drink that comes from him for our nourishment in the eternal life.
>
> *[wir seindt alle aus der erden. und die erden ist die, aus der wir gemacht seindt. nun kombt nit auch aus der erden das brot, nit auch der wein? ja; so komben sie ie auch aus dem der mensch kombt. nit daß der mensch die erden materialisch esse in irer substanz, sondern die speis, die von ir gehet, die essen wir, und das trank, das von ir geet, das trinken wir. also ist das die göttliche ordnung. so mügent ir wol auf das ermessen, daß Christus der ist, aus dem wir neu geboren werden in die ander geburt. dann aus dem seindt wir,*

> *aus dem werden wir geboren. so wir nun aus gott geboren werden, das ist aus dem leib Christi, aus seinen beinen und fleischen, so ist er auch der acker, der die frucht gibet desselbigen leibs, so aus ihm kombt, das ist die nahrung. aus dem folgt nun, daß wir nit Christum essen in der person, sonder: vom ihm die speis und drank zu unser nahrung in das ewig leben. zugleich wie von der erden speis und trank, also da auch.]*

Paracelsus clearly ties his cosmology and anthropology (concept of the composition of the human being) to his sacramental thought; in fact, he argues in the *Astronomia Magna* that the Christian receives the immortal body in the sacrament of baptism:

> For in the flesh that we receive from the Spirit we will see Christ our savior; we will not rise again and enter the kingdom of heaven in mortal flesh, rather in living flesh. However, he who is not baptized and made incarnate by the Holy Spirit will be damned. Thus we are compelled to receive baptism; if we do not, then we are not of eternal flesh and blood.
>
> *[dan im selbigen fleisch, das wir vom geist empfahen, werden wir sehen Christum unsern erlöser und nit im tötlichen fleisch, und in dem lebendigen fleisch werden wir auferstehen und eingehen in das reich gottes. was aber nit getauft wird und vom heiligen geist incarnirt, das selbig gehet in die verdamnus, daraus wir dan gezwungen werden, den tauf zu empfangen, wo nicht, so seind wir nicht des ewigen fleisches und bluts.]* [16]

Although, and befitting the age of the Reformation, Paracelsus wished to present himself as a Christian who grounded his thought in scripture (and thus lend authority to rather than foster suspicion of his ideas), nevertheless he obviously possessed the Gnostic impulse concerning the devaluation of the material realm. To him earthly and sidereal matter (i.e., mortal matter) was fit merely for permanent destruction and could not possibly be associated with the eternal heavenly realm. In fact, as I discuss below, Paracelsus thought that Christ Himself could not have worn the material flesh. Paracelsus obviously had to dance around the biblical warning regarding the spirit of anti-Christ, *exempli gratia*, 1 John 4:2–3: "Hereby know ye the Spirit of God: Every spirit that confesseth that Jesus Christ is come in the flesh is of God: [3] And every spirit that confesseth not that Jesus Christ is come in the flesh is not of God: and this is that spirit of antichrist, whereof ye have heard that it should come; and even now already is it in the world." So as not to exhibit this spirit of antichrist, Paracelsus found it necessary to provide humans with a noncorporeal resurrection flesh and Christ too with a special type of flesh. Therefore, Paracelsus created a new type of flesh and in his volumes of biblical exegesis made great efforts to ground his thought in scripture.

A still unpublished text by Paracelsus on the Psalms helps one to glean a little more insight into his exegetical practice and motives.[17] Amazingly, in this Old Testament exegesis, Paracelsus advances his theory that *during the earthly lifetime* the Christian possesses both a physical body and an immortal resurrection body, and that Christ possessed only the immortal body (and not the human physical body) while on earth. In *De Cena Domini, Ex Psalterio* (1530) Paracelsus discusses and interprets a variety of Psalms, but mostly Psalms 78, 79, 80, and 115 from the Vulgate.[18] (He includes the Homilies, and these chapters correspond to 79–81, and 116 in the King James and other standard versions.) Copied in the 1560s and housed at Heidelberg, the tract *Concerning the Lord's Supper from the Psalms* reveals that Paracelsus' principal aim in interpreting the Psalms—at least in the case of this particular piece—is simply to promote his concept of the two fleshes. Whereas most other commentators, from St. Augustine to John Wesley, focus on such themes as the Psalmists' solemn thanksgivings for deliverance from extreme perils, promises to praise God publicly, and prophecies concerning Christ and redemption, Paracelsus hopes to unearth evidence from the Old Testament to advance the two-flesh concept of his "*untödliche philosophei.*"

This "immortal philosophy" of the self-proclaimed Doctor of the Holy Scriptures advances his two-creation theory, wherein God the Father created the mortal world and body, and God the Son the immortal world and body; they are two analogous universes. Paracelsus is explicit: Psalm 78 (79 in most versions), he writes, is about the *zweyerley fleysch*, the two types of flesh, natural and heavenly.[19] With regard to the natural, Paracelsus teaches that the mortal world—the Father's creation—includes all physical being, including human flesh, as well as all that is composed of subtle "star dust." Human beings themselves have a sidereal body, also called the mortal spirit, which accounts for mind and sensation. In addition, humans possess an immortal soul, which is the very breath of God, the *spiritus vitae* of Genesis 2:7.[20]

The need for a second type of flesh emerges in part because Paracelsus, unlike most, simply could not accept the notion that the mortal body of Christians would be resurrected into a state of eternal bliss. Even mind itself, which Paracelsus associates with the sidereal body, disperses upon death and returns to the place of its origin. Ashes to ashes, dust to dust, star dust to the stars. Clearly, despite evoking Neoplatonist terminology, he refused to share the view that the *mens*, the mind, is associated with immortal spirit. Thus, humans lose irrevocably their mortal flesh and mortal spirit. In a stretch of the imagination, Paracelsus believes that the scriptures reveal the irrevocably lost status of earthly corporeality in Psalms 78 (79), where "The dead bodies of Your servants . . . have been given *as* food for the birds of the heavens [and] have become . . . a scorn and derision."[21] And he keys in on the Psalmist's request that God "Preserve those who are appointed to die" so that "[w]e, Your people and sheep of Your pasture, Will give You thanks forever; . . . "[22] Paracelsus explicitly construes this Psalm, as well as Psalm 79 (80), to mean that the scorned flesh is utterly and enduringly destroyed but that a new flesh will emerge to provide for the everlasting preservation of God's people.[23]

He adds in other Eucharistic tracts that a related passage occurs in Job 19:26, which reads, "And though after my skin worms destroy this body, yet in my flesh shall I see God."[24] The body is devoured, and yet the scriptures clearly proclaim a bodily resurrection, a reunification of body and spirit. Paracelsus achieves a bodily resurrection, *id est* a reunification of the immortal soul and body, with his interpretation of the "new creation" discussed by Paul in 2 Corinthians 5. As Paracelsus explains, Christ Himself created a spiritual or angelic body analogous to the natural body; this new creation accounts for the resurrection body, but it is also the spiritual food of the Eucharist, and the very body of Christ Himself.[25]

This approach to the composition of humans is typical of Paracelsus' manifold Eucharistic tracts, wherein he grounds his sacramental thought in his idiosyncratic, but literal, interpretations of such biblical passages as 1 Corinthians 15 and John 6:27. The latter verse reads: "Labour not for the meat which perisheth, but for that meat which endureth unto everlasting life, which the Son of man shall give unto you: for him hath God the Father sealed."[26] Paracelsus takes this to mean that Christ gives humans, via the sacraments of baptism and the Eucharist, an immortal flesh; during one's natural life this new flesh coexists with the mortal flesh, mortal spirit, and soul. He also evokes Matthew 26:26, in this context: "And as they were eating, Jesus took bread, and blessed it, and brake it, and gave it to the disciples, and said, Take, eat; this is my body."[27] The inclusion of this verse seems odd to many readers, but it is crucial to Paracelsus' views on the creation of the individual eternal bodies. This is because he parallels Matthew 26:26 with the creation of the first Adam: in the first creation, God the Father had fashioned the human out of clay (which he calls *laymen* or *limbus*); in the second creation God the Son made new creatures from his body and blood. In the first case, the Father had combined speech (a *fiat*) with clay to create a human; in the latter case, Christ spoke the words, "*accipite et comedite hoc est corpus meum*" in order to generate the second creation. These words were combined with his blood to form the "limbus of the New Testament." Paracelsus writes that when Christ broke the bread, he broke it into many parts, for "he did not wish to make only one human, rather many . . . For God the Father made only one human."[28] Thus, here we find the origin of the individual eternal bodies of humans.

Having already presented a peculiar exegesis that seemingly would have offended the sensibilities of the vast majority of Catholics and Protestants alike, Paracelsus then takes a docetic and/or monophysite turn when it comes to Christology. He evokes the angelic body—the heavenly flesh—to account for the body of Christ. Paracelsus refused to accept the notion that Christ was born through a human body dirtied by the fallen Adamic body. Actually, Paracelsus reasons, to be God's mother, it is not enough to be either a virgin or even a daughter of a virgin (Anna)—Adam's seed is genealogically defiled and Mary can in no manner be related to it. (And here one can see vestiges of Paracelsus' early Mariology; he is clearly influenced by the idea of the "immaculate conception," wherein even Mary—in order to be untainted by original sin and, therefore, acceptable for carrying God in her womb—was born of a virgin. But Paracelsus takes this further.)[29] He goes so far as to write: "*Gott hat in das todtlich fleisch nit geheurat*,"[30] that is, God did not get *that* intimate with a woman of fallen-mortal flesh. Whether or not preoccupied with sexual ramifications, Paracelsus clearly believed the Virgin had to have a body that did not descend from the line of Adam and Eve. In his early speculations of 1524, incidentally, Paracelsus goes so far as to place Mary in the Godhead—as Michael Bunners relates, "So steht die Jungfrau gleich neben Gott, . . . sie war heilig und selig vor dem Sohn."[31] Although Paracelsus would moderate his views on Mary, he consistently taught that Mary received a body fashioned by Christ in the new creation, and thus possessed a body fit for God's birth. Also pervasive throughout Paracelsus' religious oeuvre is the idea that Christ was not clothed in the elemental or sidereal bodies, but only the special type of body. Paracelsus did not publish such ideas, but they abound in his theological works and can be seen even in less well-known parts (especially Book II) of his magnum opus, *Astronomia Magna*, first published and edited in 1571 by Toxites. Again, Paracelsus is proffering a denial of Christ's humanity (e.g., monophysitism) or the argument that Christ appeared only like a human (docetism). Orthodox Christians have long insisted that Christ is both fully divine and fully human, and they point out the necessity of this for salvation. The "condign satisfaction" depends on the fact that Christ suffered and was tempted within and by the same flesh that humans possess.

It is no wonder that few early moderns would or could articulate Paracelsus' theology. The Paracelsian Alexander von Suchten implied in his *Chymische Schriften* that Paracelsus taught that the physical body would resurrect.[32] Toxites, in the prefaces to Paracelsus' texts that he edited. hinted that he accepted Paracelsus' concept of the resurrection body, but note that he evades the idea that Christ possessed a special, non-Adamic flesh:

> [T]he flesh of Adam may not go to God, for it is of the creation and is bound to death, and again must become that which it was, as the Scripture witnesses: "You are dust and will return to dust." [Gen. 3:19] And Paul adds, "The flesh and blood will not possess the kingdom of God." [1 Cor. 15:50] And, however, the holy Job says that he, [while] in his flesh, will see God his savior. [Job 19:26] Thus we come to the second flesh given to the human in the resurrection. For the human must

come into heaven not as a spirit, but as a human in flesh and blood; in this way he may be distinguished from the angels. Thus the flesh of the new birth must be there, not the flesh that the worms eat. And Christ himself corroborates this in John 3, when he says: "It is then that unless one is new born, he can not see the Kingdom of God." [John 3:3] And were we to depend (*anhangen*) on this new birth, we would make everything possible through the spirit of Christ. However, we do the opposite and hold onto the old birth; therefore we have little potential. This new birth in Christ should be well considered, and Theophrastus [Paracelsus] himself industriously describes it.

[... /nachdem das Fleisch von Adam her zu Gott nit mag kommen/dieweil es von der Creatur/vnd dem Todt vnderworffen/vnd wider das werden muß/ das es vor gewesen ist/wie die Geschrifft bezeuget: Du bist staub/vnnd zu staub soltu werden. Vnd auch Paulus sagt: Daß Fleisch vnnd Blut das Reich Gottes nicht werden besitzen. Vnd aber der heilig Job sagt/daß er in seinem Fleisch sehen werde Gott seinen Erlöser/So folgt ja das ein ander Fleisch dem Menschen geben wirdt in der aufferstehung. Dann der Mensch nit als ein Geist/sonder als ein Mensch in Fleisch vnd Blut gen himmel kommen muß/ damit er von den Englen ein vnderschiedt habe. So muß nit das Fleisch/ das die Würm fressen/sonder ein anders/nemlich das Fleisch der neuwen geburt da seyn/welches vns der Hern Christus geben wirdt/daß wir Fleisch von seinem Fleisch/vnd Bein von seinem Bein werden. Vnd das bestettiget Christus selber Johan. am dritten/da er sagt: Es sey dann das jemandt von neuwem geborn werde/so kan er das Reich Gottes nit sehen.][33]

It is immediately apparent that Toxites also fail to note that the reception of a new flesh via the new birth occurs while still on earth. Further studies can inquire into whether he was worried about the consequences of this theory of Christology and the presence of the resurrection body during this lifetime.

Returning to the archenemy of Paracelsianism, namely, Erastus, I note that it is clear that he had no access to the specifically theological texts. His criticisms of Paracelsus betray a startling lack of knowledge regarding Paracelsus' corpus as a whole. They show that he was too dependent on the opinions of others and possibly relied on spurious tracts that were not exemplary of the authentic Paracelsus' thought.

With his misidentification of Paracelsus' heresy, Erastus perhaps missed the opportunity to damage further Paracelsian philosophy—including its revolutionary chemical medicine and magical cosmos—and at the same time Paracelsus' followers seemed to succeed in making Paracelsus look as though he had proffered a somewhat respectable theology and a "Christian" alternative to the "paganism" associated with Aristotelian natural philosophy and Galenic medicine.[34] As Michael Michael T. Walton argues, Paracelsus blended his religion with his chemical and medical theories in order "to make chemistry and medicine truly Christian by freeing them from their pagan roots," adding that Petrus Severinus (1540–1602) attempted to make Paracelsus appear more orthodox.[35] Walton adds that many "adepts" sought a new Christian natural philosophy, and Paracelsus, the "German prophet," filled their need: "[T]he undeniable religiosity of Paracelsianism, when modified to appear more orthodox, proved able to withstand the criticisms of Erastus and to establish itself as a Christian alternative to pagan natural philosophy."[36] Allen Debus adds that some Paracelsians even taught that Paracelsus was restoring the genuine healing art known to Adam, Abraham, Moses, and Hermes Trismegistus. For example, Richard Bostocke wrote in 1585 that Paracelsus was doing "no more than Wicklife, Luther, Oecola[m] padius, Swinglius, [or] Calvin . . . when they . . . disclosed, opened and expelled the Clowdes of the Romish religion . . . [which] had darkened the trueth worde of God."[37] Another example discussed by Walton is Oswald Croll's "Praefatio Admonitoria" in *Basilica chemica* (1609), published in Frankfurt, which was a center for alchemical, Paracelsian, and radical religious publications: Croll was inspired by Paracelsus' claim that one learns true medicine from God alone, and he believed Paracelsus offered the true philosophy because of its grounding in both Christ and nature as well as its agreement with scriptural cosmogony.[38]

It was not until the twentieth century, when Michael Bunners and Kurt Goldammer analyzed Paracelsus' theology in depth, that scholars began to note the true heresy in his thought. Thus, during the early modern period, Paracelsian medicine—and its corresponding *chymistry*—immersed in a propaganda war with the university establishment, was able to win more and more adherents despite Paracelsus' Gnostic deprecation of the flesh. Although Walton believes, contra the depictions above, that "[t]he idiosyncratic nature of [Paracelsus'] religious ideas was a major impediment to the acceptance of [Paracelsian] philosophy," it seems that a much more severe impediment would have emerged if the *theologica* had been published.[39] Having evaluated the problems inherent within the accusation that Paracelsus was an Arian, I will now shift to the charge that the Hohenheimer consorted with demons.

Paracelsus on Magic

Naudé, mentioned above in the introduction, aptly captured Paracelsus' approach to magic: Paracelsus was not a practitioner of magic and should not be charged with "magick." That is, Paracelsus merely discussed the types of magic and the mechanisms behind them. Here, drawing from passages in the *Astronomia Magna*, I let Paracelsus speak for himself on the topic of magic. And I show that not only is Paracelsus' discussion of the subject of necromancy and conjuration benign, but also that he was certainly not a proponent of working with demons; he means something entirely different when presenting "demonic magic."

The best place to begin with regard to understanding Paracelsus' understanding of magic is the *Astronomia Magna*, Book I, which is entitled "The book of the philosophy of the heavenly firmament" (*Das buch der philosophei des himmlischen firmaments*). In Chapters 1–3, the most widely read, reproduced, and translated section of the *AM*, Paracelsus lays the ontological foundations of his picture of man and the cosmos, and his focus is the creation by God the Father. In Chapters 4–11, he discusses "natural astronomy" in terms of its nine divisions (*membra*) and ten gifts (*dona*). With his nine *membra*, Paracelsus uniquely classifies the types of *adept* art, for example, magic, adept medicine, nigromancy, and astrology—these can be studied and mastered. The gifts, on the other hand, are the "aethereal arts" brought about by nature herself without human aid: these include impressions (e.g., wisdom and prudence), generation (growths from the four elements), and "*inanimatum*" (e.g., nymphs and gnomes).[40] Thus, natural astronomy is basically Paracelsus' expression for natural magic, *id est* the arts and operations associated with the subtle matter of the universe that proceeds from the stars. It is important to note that he often uses the terms "magus" and "astronomer" interchangeably. Clearly, Book I of Paracelsus' magnum opus is not merely a catalogue of the types of occult arts, though it is that too; natural magic is a systematic natural science, the subject matter of which includes a wide array of natural phenomena. Thus, Book I is also a statement of Paracelsus' epistemology: mortal human intelligence is associated with the sidereal component within humans as well as the instruction given by the stars. It is also a universal cosmography: Paracelsus does not adhere to the Aristotelian-Scholastic conviction that the terrestrial and celestial realms each possess their own matter and physics. Instead, he seeks to explain the intimate interaction between the firmament and the earth, as well as the mechanisms and substances that the two hold commonly. Paracelsus writes,

> Thus, it should now be understood that, first of all, there is a body about which the *astronomus* can say nothing. It is elemental, belonging to the earthly [topics] in the elemental philosophy. But of the *natural spirit* in the body, the *astronomus* has the power and might to discuss, for the body, which is of the elements, is married (*vermälet*) to the spirit which is given to the elemental body; [the elemental body is thus] incarnated (*eingeleibt*) by the firmament. Thus, philosophy is divided into two parts, first in the essence (nature) of the spirit and second in the essence of the body, that is, into body and spirit. (My emphases.)
>
> *[also sol nun verstanden werden, das erstlich ein leib ist, von dem der astronomus nichts redet; er ist elementisch, gehört in die irdisch der elementen philosophei. aber von dem natürlichen geist im leib hat der astronomus zu reden gewalt und macht; dan der leib ist der elementen, der geist, der dem elementischen leib geben ist, der ist im vom firmament vermälet*

und eingeleibt. und also teilt sich die philosophia in zwen teil, in das wesen des geistes und zum andern in das wesen des leibs, das ist in den corpus und spiritum.][41]

Paracelsus' overall approach to magic can be best understood via his discusion of the types of adept art. Beginning with the science of "*magica*" [*sic*], Paracelsus differentiates magic from nigromancy, chiromancy, and other esoteric arts. Introducing magic in its species, Paracelsus writes:

> The six species were regarded by the Saba in the orient and those on the island of Tarsus to be the highest wisdom given by God to humans in mortal life. And only the wise men who possessed this art were called magi, and all other mortal wisdom was diminished and thought weaker, and only magic was held to be the most splendid and insuperable wisdom.
>
> *[dan dise sechs species haben die in Saba in oriente und in der insula Tarsis für die höchste weisheit geachtet, so von got dem menschen in tötlichen leben geben seind. und alein die sapientes, die solchs kunst haben, hat man magos geheißen, und weiter alle ander tötliche weisheit gemindert und schwecher gehalten, und alein die magicam für die trefflichste und unuberwintlichste weisheit gehalten.]*[42]

Paracelsus considers magic to be the interpretation of *natural* signs that God placed in heaven *supernaturally*, and his focus is on oriental magic, for example, the art of the magi who followed the star of Bethlehem. In the *Labyrinthus medicorum errantium* (1538), Paracelsus writes that all types of magic originated in the orient, and that "nothing good comes from the *septentrione*."[43] Because of its association with Eastern wise men, such as those attending Christ's birth, the magical art is called *artes sapientiae*. Nigromancy concerns the recognition of the sidereal and elemental spirits after death, as well as the spirits existing in the heavens. The nigromancer can also reach into a person's body without causing injury or hide a person by making him or her invisible.[44] Contrasting with nigromancy is *nectromantia*. There are beings, called *flaggae*, who exist throughout the world in objects of all sorts. They observe people and thus know secretive things. A nectromancer can win these *flaggae* over so that they are obedient and willing to reveal the secrets of others, such as the places where stolen goods are hidden. There are many types of nectromancy, or ways to make the *flaggae* visible, including the use of beryls and mirrors.

Turning to astrology, Paracelsus notes that an astrologer must know the function of each star, that is, what each intends to do at a given time and what each cannot do. The astrologer who knows the capabilities of the stars is better than the one who knows only their motions. The astrologer also recognizes nature's highest guide (*summum motorem*), which, much like Aristotle's prime mover, holds nature in its hand and directs the firmament completely—just as if it were a prisoner. An astrologer should know the nature, complexions, and qualities of the stars as well as a doctor knows the nature of the ill; it is also necessary to become familiar with the stars' concordances (*concordantias stellarum*), recognizing their relation to humans, animals, the four elements, and all things that grow and spring forth from the elemental mothers (*aus den matricibus elementorum*). The astrologer should be knowledgeable—with natural understanding—in every manner and form of the firmament, just as the philosopher or doctor is in natural things. "The [firmament] has all the qualities of a human, but it is so confined in the carrying out of its work that it can not do what it would like to do (*derselbig hat alle eigenschaft, die ein mensch haben sol, iedoch aber volbringunge der werken sind im verhalten, das er das nit tun mag, das er gern tete*)."[45] Again, the firmament is like a prisoner in the hand *summi motoris*; an astrologer cannot see how this highest mover directs the firmament.[46]

Regarding signatures, Paracelsus writes that nature has shown nothing that cannot be learned through signs. There are four species that contain all natural signatures in themselves. The first of these species regards the forms of the human. As the stars have their signs that may be recognized, so does the human; the difference is that the human is seen and recognized through fixed lines (*lineas fixas*) rather than through his or her movements; that is, one can study the complexions associated with celestial motion, or one can study the complexions of lines in humans. As an herb grows into the form of the nature to which it belongs, so too does the human; the form reveals what type of herb it is, and the *signatum* shows who a person is. The quality in the human is the causal factor, not the name or sex. "The art of signs teaches the correct name given to each as is born within."[47] Paracelsus means, for example, that an animal containing the form of a wolf will be called a wolf, not a sheep.

All natural secrets that lie within the human can be revealed through the four species of signatures. Chiromancy uncovers these in three ways. As these lines and veins open natural secrets, so too can physionomia, which is formed and placed according to the contents of one's senses and disposition (*gemüt*). The fabricator of nature is artistic, not melding the disposition after the form, but the form according to the disposition—that is, a person's appearance—will be determined by the innate qualities that he or she possesses. The uncertain arts, so named because they are associated with the power of imagination rather than any particular instrument, enable their practitioners to learn secrets via the tangible elements. When one imagines something, a new constellation is created in the firmament, which can then be utilized so as to glean insight into, for example, one's fortune, a person's location, where something is hidden, or even the future. One may see these in fire, water, wind patterns, or other formations. But imagination can have other effects: for instance, parents can change their child's constellation—hence talents—with their ponderings.

On adept medicine, Hohenheim writes that there are two types of medicines: one from the earth, one from the heavens. Thus, half of medicine occurs in the realm of astronomy. A doctor must understand the difference between the elemental and sidereal in order to differentiate the diseases from one another; after all, one may become infected due to astral influences.[48] When it comes to adept philosophy, Paracelsus is again primarily interested in medicine. He notes that unlike a philosopher, who can describe the natural powers in herbs, the adept philosopher is privy to the firmamental component in a plant. As medicine springs forth from the earth, so too does it come from the stars; it floats, possessing no tangible body. Therefore, an alchemist, who is a natural *philosopher* and not an *astronomer*, is not able to extract this firmamental medicine as he can a virtue from an herb—the employment of astral medicine is limited to those sagacious in astral philosophy.[49] Adept mathematics, as opposed to mathematics proper, has to do with abilities and instruments associated with the counting and measuring of bodies, that is, insofar as they relate to sidereal powers.

For a further sample of Paracelsus' approach to the various species of the divisions of natural magic, and also to show that Paracelsus was clearly not condoning an alliance with dark powers, let us examine more thoroughly his thoughts on nigromancy and nectromancy. On nigromancy (necromancy), Paracelsus writes:

> In order that you correctly understand nigromancy and the nigromancer, note that nigromancy is divided into five species. The person who can use and knows the five species can employ nigromancy and is a nigromancer, which is the second part of astronomy. And the first principle of this division is that the eternal and mortal parts of the human are divided after death: thus there are two mortal bodies that remain on earth, the elemental and sidereal components left behind by the human. Concerning the knowledge of the first species, note that a nigromancer is one who recognizes the two spirits, and possesses knowledge of their qualities, essences, and types.
>
> *[Damit ir nigromantiam und nigromanticum recht verstandent, so merkt, das nigromantia ausgeteilt wird in fünf species. welcher die fünf species kan und weiß, der kan nigromantiam und ist ein nigromanticus, und ist das ander membrum astronomiae. und das erst fundament dises membri ist also, das der mensch nach seinem tot, so das ewig und das tötlich von einander gescheiden seind, so bleiben zwen tötlich geist auf erden, so der mensch hinder im last, den elementischen und den siderischen. iezt folgt auf das die erkantnus der ersten species also. wer solche zwen geist erkent, was sie seind und weiß ir eigenschaft, wesen und art, derselbig kan die erst speciem nigromantiae, gleich also zu verstehen.]*[50]

The second type of nigromancy concerns the person who can deal with the spirits of the dead so that they serve him and improve his business. One can thus utilize the dead as one would a servant; here the servant will not eventually become his master's master. When employing this *tortura noctis*, one should observe the senses the spirit held during life. The *metheorica vivens* regards the many types of floating spirits born in the stars that exist and die in the chaos. Those who recognize and call these spirits to service—in the same manner as a doctor utilizes herbs—have learned the third type of nigromancy. The person competent in *clausura nigromantica* can reach into a human body without injuring the person, just as if they had reached into water to pull out a fish without causing a hole in the water. Reaching into a person, this type of nigromancer can thus do such things as pull items out of someone or place something inside the body. The fifth type, *obcaecatio nigromantica*, occurs when a visible body is made invisible. One can thus hide things just as the darkness of the night does.

Concerning nectromancy, there are two general categories. The first pertains to making the *flaggae* visible. In the second case, the *flaggae* implement the will of the nectromancer invisibly. Everything that a person says is already in the mouth of another, and all a person's deeds and effects are manifest. Secrets about prices, for example, may be betrayed—the *flaggae* are ever watchful. Through the art of nectromancy the *flaggae* must obey those who make them visible. This can happen through, for instance, mirrors, glass, or coal. They are compelled to make visible not only themselves, but also that which they know that is hidden. If the *flagga* does not become visible through the first species, then one can employ one of the others, for example, compelling the *flagga* to use an object with which to point in a certain direction or arranging objects in telling figures. Through this art one can find hidden treasures and read confidential letters. A complete nectromancer can make a *flagga* entirely obedient. Nectromancy functions through both rewards and violence. The *flaggae* can thus be compelled to become visible in such items as mirrors, beryl, and coal. They can be forced to show and illustrate things via the use of thickets, lead, stones, and so on. They can also extinguish candles and reveal secrets in similar ways.[51] Paracelsus then presents a fascinating nigromantic proof.

First, Paracelsus reminds the reader that the elemental body is tangible, and the sidereal body intangible. He continues to reiterate that the sidereal body from the stars exists in the air after death, and here the stars will consume it. As the elemental body takes some time to decay, the sidereal body does too. In the meantime, the two bodies— now separated—retain their old movements, manners, and conduct, remaining in the vicinity of where they had lived, the elemental in the grave, and the sidereal floating in the air, seeking the place where it had dwelled in life. The sidereal body can thus be seen. When a person's spirit is seen, notes Paracelsus, this is the sidereal body, which is like an image in a mirror; it is neither a human nor soul. Of course, the person talented in nigromancy knows well that the sidereal bodies are able to reveal treasures and hidden places.

Paracelsus then presents his explanation for the art of nigromancy, explicitly calling it a natural process. All sorts of errors arise from mistaking these bodies for souls, like intercession, *exempli gratia*, calling on saints in heaven to pray for those on earth—such is fostered by the Church and its corrupt leaders. Characteristically, Paracelsus has thus offered a naturalistic argument against Church ceremony and "superstition." He even goes on to deride those who believe that the "person" (really not a person, but the sidereal body) who has been encountered sits in hell or purgatory, where he or she does penance until the Day of Judgment. In contrast, he reflects, one is simply either saved or damned. Paracelsus considers the idea of speaking with souls in purgatory foolish, and chides those who consider such "fantasy" to be great wisdom: one cannot speak with souls and purgatory is an invention of liars.

It follows that false orders arise, such as those of the exorcists and conjurers. Dead things cannot speak, insists Paracelsus, so the exorcists are speaking with no one. The conjurers are merely calling forth sidereal bodies (mortal spirits). The result of such thoughtlessness is that the devil possesses the sidereal spirit, which has been mistaken for a conjured "soul," and Satan takes satisfaction in their recklessness. Of course, the devil can possess living humans, but it is still easier to enter the dead spirit (sidereal body), wherein there is no resistance. It follows that these conjurers are dealing with the devil, not the dead, and their joy is demonic. Furthermore, Paracelsus fumes, those adherents to the books of the dead are those who attempt to receive intercession from those in heaven, hence praying to the sidereal spirits. He believes that they forget that the human's (i.e., the soul's) dwelling place after death is *not* in the grave (where the elemental body goes), and not in the air (where the sidereal body may be seen), but rather that he has another place, where he is held until the time of probing, that is, when the trumpet sounds and it is spoken, "Arise dead ones."[52]

Having listened to Paracelsus explain his concept of magic and reject the idea of conjuring dead souls, let us turn to Book IV of his *Meisterstück* in order to see precisely what he means by "demonic magic":

> In order that infernal magic may be understood, as well as other species of astronomy, it is for this reason that I undertake the task to elucidate these things, how they work—*id est* the functions specific to their natures—so that we know what demonic impressions are and what is not demonic, and so that [we also know], however, that all [natural and infernal astronomy] is indeed *natural*, and is revealed to be fundamentally natural. . . . So therefore, reader, read the following on the astronomical philosophy of the infernal crafts. In this way you may

> recognize and know the infernal strengths and powers, and thus protect yourselves from these in your dealings, and protect yourselves from error.
>
> *[Damit das magica infernalium verstanden werde oder ander species der astonomei, aus der ursachen ist mein fürnemen, dise ding zu eröfnen, wie sie wirken, wie sie hantlen in irem wesen, damit wir wissen, was teuflische impressiones seind, was nicht teuflisch, und aber iedoch alles natürlich, und wie natürlich, grüntlich angezeigt werden. . . . darumb leser so lis nachfolgents die astronomische philosophei der infernalischen kreften, auf das du mögest erkennen und wissen, was infernalische kreft und macht sein, und dem selbigen nach hantlen und sich vor irtumb behüten.]*[53]

Of import for the topic of infernal astronomy, Paracelsus notes that fallen angels have not lost their intelligence and talents, only their position:

> For it is of great importance to observe God's created ones (*geschöpf*), namely, the angels who are gifted with great wisdom, which humans do not possess so highly. So, now the devil was Lucifer, and although he lost [his battle in heaven] the same, he has, nevertheless, merely lost his place, that is, [his presence in] the countenance of God, but his kingdom, art, and wisdom remain with him, for *scripture does not report* of his being deprived or relieved of these. (My emphasis.)
>
> *[dan es ist ein großes, zu betrachten die geschöpf gottes, nemlich die engel mit großer weisheit begabt, den menschen so hoch nicht. so nun der teufel Lucifer gewesen ist, ob ers gleichwol verloren hat, so hat er doch nur die statt verloren, das angesicht gottes, und sein reich, kunst, weisheit ist im bliben; dan die geschrift meldet der selbigen beraubung oder entsezung nit.]*[54]

"Evil spirits" can thus deceptively employ all things "*anatomias, secreta, arcana* und alle große *magnalia dei* und *mysteria naturae*."[55] They deceive doctors so that they murder people with false medicine.[56] They also bring inclement weather, such as thunder, hail, snow, wind, and rain. To account for their meteorological talents, Paracelsus reaches into his philosophy of meteors (*meteorische philosophei*) and theory of the elemental mothers. The firmamental impressions are born from three things: salt, sulphur, and mercury. The archeus of the firmament forges and prepares the impressions (fruits) from these three. The devil has the same abilities, and would make life impossible for humans if given free reign to use the firmamental seeds to fashion hail, storms, dangerous weather, and strange heavenly signs whenever he wanted. God allows him to wreak havoc occasionally, and the result always portends future ills: the devil may only forge signs that reveal menacing omens. Clearly, those who wish to make predictions on the basis of the weather and heavenly signs must be able to differentiate between the devil's doings and those of God.[57] Paracelsus believes that infernal astronomy is thus a useful art, but that does not mean that he advocates being in league with demons.

Addressing nature in an ethical and theological manner, which includes his thoughts on the reason for life in the mortal realm, Paracelsus writes that all humans, animals, and things of the elements and firmament are divided into two types: good and evil. This appears to be useful to God, for before he can invite anyone into his kingdom, it is necessary to put them through a trial, like the gold in fire (*wie das golt im feur*).[58] As the angels were tried, so too must humans be. Thus, both angels and humans were given free will, so that God could observe their choices under duress.[59] Characteristically, he takes the opportunity to lambaste the powerful and wealthy by drawing an analogy with the Fall of Lucifer (at which time he took on an ugly form): "[God] expelled the powerful from their seat (that is, Lucifer as the most powerful angel) and elevated the humble, that is, he gave another position to the meek and poor angels. This proves that here on earth wealth is against God."[60] He concludes that those who deviate from the path of the Lord become possessed, becoming an evil spirit; he offers the examples of the fraudulent businessmen and the miserly, who he despises as much as the thieves, drunkards, liars, and prostitutes.[61]

Given Paracelsus' clear advocacy of Christian ethics in his writings on magic, one wonders why the accusation of "consorting with demons" gained any traction at all. This attack, alongside the charge of "Arianism," was part of an ill-conceived effort to soil Paracelsus' legacy. The detractors of Paracelsus simply misunderstood and/or misrepresented his thought, and was this the case with his proponents as well?

Conclusion: The Incomplete and Misunderstood Reception of Paracelsus

As noted above, the Paracelsians embraced Paracelsus' ideas pertaining to magic, but even they seemed mostly unaware of the true nature of his theological heresy. The only early modern publication of a Eucharistic tract by Paracelsus was the *Philosophia de Limbo, aeterno perpetuoque homine novo secundae Creationis ex Iesu Christo Dei Filio*, which appeared in 1618 alongside a few of Paracelsus' other religious tracts.[62] These works were part of a larger collection published by Joannes Stariz in Magdeburg. Here Paracelsus appears in a fascinating and heterodox assortment of works by authors who seem to be united only in their eccentricity, and the Eucharistic philosophy of Paracelsus remained an obscure topic.

Perhaps most telling with regard to the reception of Paracelsus' theology is a popular text attributed to Paracelsus that saw several publications between 1605 and 1736, namely, *Kleine Hand und Denckbibel*. Astonishingly, the introduction to this Paracelsus Hand Bible portrays a very docile and orthodox-sounding Paracelsus, and as the dean of Paracelsus studies, Karl

Sudhoff noted, in his 1894 *Kritik der Echtheit der Paracelsischen Schriften*, the text is by someone who does little more than select and present verses from the Old and New Testaments. Sudhoff adds that not a word is taken from Paracelsus' numerous theological manuscripts, and nowhere does one find a trace of the authentic Paracelsus.[63] One might ask: Was this work simply propaganda, that is, an effort to present Paracelsus as a good Christian?

Paracelsus' legacy was simply not tainted by the Gnostic-Docetic-Monophysite blemish. Either no one cared enough about the exceptionally controversial theological ramifications to give his theology its due attention, or his more controversial ideas simply remained safely obscure. As mentioned earlier, the historian Carlos Gilly goes so far as to point out the sidestepping by editors Michael Schütz (Toxites) and Adam von Bodenstein of "specific theological issues in the prefaces to their editions of Paracelsus."[64] Paracelsus was controversial enough as it was, notes Gilly, and thus theological Paracelsianism necessarily experienced an "enforced clandestinity."[65] Why would one present Paracelsus' docetism and unsavory picture of the resurrection of the saints when attempting to promote the chemical pharmaceutics, pioneering pathology, and revolutionary matter theory of the iatrochemists? The High Priest of the chemical philosophy—as Allen Debus dubbed Paracelsus—required a more polished image and already faced enough acute criticism. Further questions emerge in this context: Did Paracelsus' volumes of religious writings, including the exegetical tracts advancing his Eucharistic philosophy, serve any important role in the advancement of his new approach to medicine and understanding of the cosmos? Does the centuries-long lack of acute attention to his biblical studies betray the fact that his theology has usually been seen as embarrassing and heretical? Why did any early moderns work with or publish the explicitly theological tracts at all? Why are heretical notions related to Christology notably absent as is the case in nearly every printed work attributed to Paracelsus?

The assessment of Erastus' vitriol against Paracelsus shows that the Heidelberger was incorrect—and probably misguided, even intentionally misleading—in his assertions regarding Paracelsus' magic and Christology. Concerning magic, not only was Erastus wrong about Paracelsus' relationship to demons, but also that the aspects of Renaissance magic and hylozoist matter theory that resounded in Paracelsus' thought were not obstacles to the spread of his ideas, but rather helped promote it among the antiestablishment and esoteric types. After all, many did not consider magic to be incompatible with "proper" Christianity. One may note the popularity of others who wedded the two, for example, medieval alchemists such as pseudo-Lull and Renaissance thinkers like Marsilio Ficino. (In addition, it is conspicuous that even Paracelsus' popular *Denckbibel* is published together with the Paracelsian tract, *Untersuchung des Glaubens, su samt dem Haupt-Schlüssel der Paracelsischen Arcanen, für die Liebhaber aufs Neue vermehret, und zum Druck befördert.* In other words, the *arcana* of Paracelsians—related to magical powers—were obviously considered by many to be "Christian," that is, perfectly compatible with the biblical passages next to which they were published.)

Concerning Paracelsus' theology (utterly misunderstood and misrepresented by Erastus), his reputation as a heretic never really gained traction because it was conveniently overlooked or covered up. The unpalatable aspects of Paracelsus' theology were met with a denial of and/or lack of interest in them. The two-Creation cosmos of Paracelsus, exceptionally important to him and helpful to us with regard to understanding his thought, seems to have had very little impact on his followers. His theological heresy was mostly unknown in early modernity, and the extent of its influence on his most vehement proponents—for example, Bodenstein, Suchten, and Toxites—and Paracelsianism as a whole requires further study. Regardless of the continuing confusion and questions surrounding Paracelsus and his reception, unlike the early moderns we have the benefit of access to most of his oeuvre—Paracelsus' cosmos becomes much more intelligible when we recognize the extent and importance of his unique two-Creation theory within his overall *Weltanschauung*, and when we discern the difference between Paracelsus' approach to magic and his detractors' rhetoric about his magic.

Notes

I would like to thank the following: William Newman, Andrew Weeks, Hartmut Rudolph, and Jole Shackelford provided indispensable support with regard to the research and writing that went into this work; and Barbara Williams [Archivist, Health Sciences Campuses of Drexel University (Hahnemann Collection and holdings of the Medical College of Pennsylvania (MCP), formerly Woman's Medical College)] was exceptional in helping me with the Hering Paracelsus Collection. I am also grateful to the Chemical Heritage Foundation in Philadelphia as well as Heinz Schott and the Institute for the History of Medicine at the University of Bonn, both for financial support and access to their collections, and Wright State University, Lake Campus, for a Faculty Development Grant that helped provide travel support to Philadelphia and Bonn, Germany. Furthermore, I wish to express my thanks to the University of Heidelberg and Leiden University for their assistance and use of their archives.

1. Gabriel Naudaeus, The History of Magick By way of Apology, For all the Wise Men who have unjustly been reputed Magicians, from the Creation, to the present Age, trans. from the French by J. Davies (London: Printed for John Streater, and are to be sold by the Book-Sellers of London, 1657), 187–88: "But, however it be, I shall not quarrell with their opinion, who

hold, that one of the principall advantages which learned and industrious men have over the ignorant, is, that it is in their power to make new Systems, and advance new Principles, nay change the order, precepts and method of the Sciences, shortening or lengthening them, like a Stirrop, as they please. Of which number Paracelsus being one thought he might as well inverst the course of Magick, as he had done that of Medicine and Philosophy, and boasted he could have done of Religion, threatening both the Pope and Luther to bring them both to his Maxims when he should think fit to do it. Though therefore he might justly be condemn'd as an Arch-heretick for the depravednesse of his opinion in point of Religion, yet do I not think he should be charg'd with Magick. For this consists not in the Speculations and Theory; which every one may explicate and amplify according [188] to his fantasie, but in the practice of the Circle and Invocations, wherein, as we have already shewn, not any one of the Authours, that have the greatest aversion for his Doctrine, would ever maintain he employ'd himself."

2. On Erastus' anti-Paracelsianism, see Charles D. Gunnoe, Jr., "Erastus and Paracelsianism: Theological Motifs in Thomas Erastus' Rejection of Paracelsian Natural Philosophy," in Allen G. Debus and Michael T. Walton, eds. *Reading the Book of Nature. The Other Side of the Scientific Revolution* (Kirksville, Missouri: Sixteenth Century Journal, 1998), 45–66; 48; Bruce T. Moran, "Paracelsus, Religion, and Dissent: The Case of Philipp Homagius and Georg Zimmermann," *Ambix* 43 (1996): 65–79; and Carlos Gilly, "Theophrastia Sancta: Der Paracelsismus als Religion im Streit mit den offiziellen Kirchen," in Joachim Telle, ed. *Analecta Paracelsica: Studien zum Nachleben Theophrast von Hohenheims im deutschen Kulturgebiet der frühen Neuzeit*, Heidelberger Studien zur Naturkunde der frühen Neuzeit, vol. 4 (Stuttgart: Steiner, 1994), 425–488.
3. Erastus, the dogmatic polemicist, who was deliberately seeking out contentious elements among the Paracelsians, would no doubt have liked to have served his agenda by misrepresenting, misappropriating, and/or overlooking any number of concepts by Paracelsus and the Paracelsians. Like his contemporaries, however, he believed the *Philosophia ad Athenienses* (considered by Karl Sudhoff, the dean of Paracelsus studies, to be of dubious authenticity) to be an authentic writing by Paracelsus, and he saw in it the Arian heresy. See Gunnoe, "Erastus and Paracelsianism," 52, n. 32 and 56, for a full discussion of both the authenticity of the text and Erastus' detection of Arianism therein.
4. A thorough presentation on the whole of Paracelsus' cosmogony would necessarily include a discussion regarding definitions of body, soul, and spirit during the fifteenth and sixteenth centuries (including polemical demonology). Elsewhere I discuss Paracelsus' views on the cosmogony of the natural world—including Stoic and Renaissance-Neoplatonic elements, and some parallels with Ficino—in Dane T. Daniel, "Invisible Wombs: Rethinking Paracelsus' Concept of Body and Matter," *Ambix* (2006): 129–142; and Dane T. Daniel, "Paracelsus' Astronomia Magna (1537/38): Bible-Based Science and the Religious Roots of the Scientific Revolution," Ph.D. dissertation

(Indiana University, Department of History and Philosophy of Science, 2003). Much of the discussion below on demonic magic is derived from Chapters 3 and 4 of this dissertation, but—again—I omit here an in-depth treatment of Paracelsus' concepts on the "mortal" components of humans, for example, the physical body and sidereal body (the latter of which Paracelsus also called "spirit").

5. One reason that Paracelsus' religious views were mostly unknown is that half of his explicitly theological (and prodigious) oeuvre remains unpublished to this day. Even the vast majority of his medical, scientific, and philosophical tracts were not gathered and published until a few decades after his death. Strangely, despite the fact that Paracelsus outlined his theology in Book II of his magnum opus, the *Astronomia Magna*, few seemed to recognize the more contentious aspects of his exegeses. For more discussion on this matter, see Dane T. Daniel, "Coping with Heresy: Suchten, Toxites, and the Early Reception of Paracelsus's Theology," in Lawrence M. Principe, ed. *Chymists and Chymistry: Studies in the History of Alchemy and Early Modern Chemistry* (Sagamore Beach: Chemical Heritage Foundation *and* Science History, a division of Watson Publishing International LLC, 2007), 53–55.
6. Theophrastus Paracelsus, *Kleine Hand und Denckbibel, oder Einführung zu der geheimen Weisheit und verborgenen Wahrheit des Geistes Gottes und unsers Herrn Jesu Christi, Nebst einem sonderbaren Tractat, genannt, Untersuchung des Glaubens, su samt dem Haupt-Schlüssel der Paracelsischen Arcanen, für die Liebhaber aufs Neue vermehret, und zum Druck befördert* (Mühlhausen: Bey Christoph Friderici, 1736).
7. Carlos Gilly, "'*Theophrastia Sancta*'—Paracelsianism as a Religion," in Ole Peter Grell, ed. *Paracelsus: The Man and His Reputation, His Ideas and Their Transformation* (Leiden, Boston, Köln: Brill, 1998), 151–186; 158.
8. Book I, Chapters 1–3 became the portion of the *AM* most often reproduced and circulated. The early chapters reached a wide readership in the popular full editions of Toxites and Huser, copies of which survive in numerous libraries throughout Europe and North America. Indicative of the special interest given to the consummate expression of Paracelsus' cosmology and magic in Book 1, Chapters 1–3 are a number of incomplete versions of the *AM* that consist of these chapters exclusively. See Karl Sudhoff, *Versuch einer Kritik der echtheit der Paracelsischen Schriften* (Berlin: G. Riemer, 1894–1899), 136–137. One is a manuscript held in Erlangen: copied in 1590 after printed versions were already available, it includes no prefaces, but nevertheless a complete copy of the first three chapters of the Toxites edition. Sudhoff inventories the manuscript as No. 43: "Erlangen, Univ. Bibliothek, Ms. 1714." Sudhoff quotes a passage from folios 405b to 448b: "Extract Ausem Opvus des Buechs Astronomia Magna: Oder Die gantze Philosophia sagax der grossen vnd kleinen Welt, des Von Gott Hocherleichten, erfahrnen, vnd bewerten teitschen Philosophi vnd Medici Philippi Theophrasti Bombast, genannt Paracelsi magni Darinn er lehret es gantzen natürlichen....... Sigismundus feyerabent Anno 1571 Mit. Rom Keyß. Mayst: gnad vnd

freyheit auf 10 Jahr *u* [...] Getruckht zu Franckhfort am Mayn, beim Martin Lechler, in Verlegung Hieronymi feyerabends *u* 1571." Another work, which draws attention to the non-German interest in the topic, is a 1619 Dutch translation of the first three chapters. See Sudhoff, *Bibliographica Paracelsica*, 519–520, #310: *Vande heymlijcheden der scheppinge aller dingen DOOR Philippus Theophrastus Paracelsus... Met noch drie vervolgende Capittelen vande Phisophie* [*sic*] *des hemelschen Firmament getrocken wt de Astronomia Magna ofte Philosophie Sagax vanden selven Autheur*, trans. M. Henrick Janssz (Leyden: Daniel Roels Boecvercooper, 1619).

9. See, for example, Michael Toxites, "Vorrede," in Paracelsus, *Astronomia Magna, oder, Die gantze philosophia sagax der grossen und kleinen Welt...* (Franckfurt am Mayn: Hieronymi Feyerabends, 1571); and C. Schmidt, *Michael Schütz, genannt Toxites. Leben eines Humanisten und Arztes aus dem 16. Jahrhundert* (Strassburg: C.F. Schmidt's Universitäts-Buchhandlung, Friedrich Bull, 1888).
10. Regarding the Weigelians' attention to Paracelsus, see Aurelij Philippi Theophrasti Paracelsi ab Hohenhaim, *Philosophia De Limbo Æterno Perpetvoq[ue] Homine Novo Secvndæ creationis ex Iesu Christo Dei Filio Des Edlen/ Hochthewren Philosophi...*, ed. Joannem Staricius (Magdeburg: Johann Francke, 1618), and Carlos Gilly, "'*Theophrastia Sancta*'—Paracelsianism as a Religion." See also H.U. Preu, Die Theologie des Theophrastus P. von Hohenheim in *Auszügen aus seinen Schriften dargestellt* (Berlin: 1839); Kurt Goldammer, "Paracelsische Eschatologie," in *Paracelsus in Neuen Horizonten: Gesammelte Aufsätze* (Wien: Verband der wissenschaftlichen Gesellschaften Österreichs, Verlag, 1986), 87–122; Hartmut Rudolph, "Hohenheim's Anthropology in the Light of his Writings on the Eucharist," in Grell, *Paracelsus*, 187–206; and Michael Bunners, *Die Abendmahlsschriften und des medizinisch-naturphilosophische Werk des Paracelsus*, Ph.D. dissertation, Berlin/GDR, 1961.
11. Charles D. Gunnoe, Jr., "Thomas Erastus and His Circle of Anti-Paracelsians," in Joachim Telle, *Analecta Paracelsica* (Stuttgart: Franz Steiner, 1994), 135. See also Gilly, "'*Theophrastia Sancta*'—Paracelsianism as a Religion," 154–155, on Gessner's belief that Paracelsus was an Arian, and also on the shared opinion by Johannes Crato von Crafftheim and others.
12. Gunnoe, "Erastus and Paracelsianism," 56.
13. See n. 3 above.
14. For a biblical verse on the final conflagration, see 2 Peter 3:11–12: "Seeing then that all these things shall be dissolved, what manner of persons ought ye to be in all holy conversation and godliness, [12] Looking for and hasting unto the coming of the day of God, wherein the heavens being on fire shall be dissolved, and the elements shall melt with fervent heat?" I will be utilizing the King James Version of the Bible throughout this chapter.
15. This translation is from my discussion of the topic in Dane Thor Daniel, "Paracelsus' *Declaratio* on the Lord's Supper. A Summary with Remarks on the Term *Limbus*," *Nova Acta Paracelsica* N.F. 16 (2002): 141–162,

158–159: Paracelsus' philosophy of the eternal limbus—presented in such works as the *Abendmahlsschriften* of the early-mid 1530s and its component, the *Declaratio*—reveal Paracelsus' tendency to inform his biblical exegesis with naturalism even as he evokes scripture to explain important elements within his natural philosophy. One can readily see an example of Paracelsus' efforts to understand the eternal according to analogies with the natural in this revealing passage from *De Genealogia Christi*, in which Paracelsus directly discusses the two types of soils. The passage is found in II, 3:81—The modern edition of Paracelsus' collected works is divided into two parts; the first division is Paracelsus, *Sämtliche Werk, I. Abteilung: Medizinische, naturwissenschaftliche und philosophische Schriften*, ed. Karl Sudhoff, 14 vols. (Munich and Berlin: R. Oldenbourg, 1922–1933). The first volume of the second division is Paracelsus, *Sämtliche Werke, II. Abteilung: Die theologischen und religionsphilosophischen Schriften, Erster Band, Philosophia magna* I, ed. Wilhelm Matthiessen (Munich: Otto Wilhelm Barth, 1923). Kurt Goldammer directed the publication of several more volumes of the second division between 1955 and 1986. In this chapter I am referring to Paracelsus' works using the Roman numeral to stand for the division (I or II) and the volume and page number will follow, for example, II, 3:81.

16. I, 12:310.
17. It is important to note that Paracelsus wrote extensively on Psalms outside the context of his Eucharistic thought. Two volumes of Psalms exegesis appear in Paracelsus, *Sämtliche Werke*, II. Abteilung: *Theologische und Religionsphilosophische Schriften*, ed. Kurt Goldammer, 6 vols. (Wiesbaden: Franz Steiner Verlag GmbH, 1955–1986).
18. Paracelsus, Ex Psalterio Declaratio Coene dominj Theophrasti Liber, in Codices Palatini germanici 26, fols. 198a–211a, Universitätsbibliothek Heidelberg.
19. Ibid., 211b: "Ex Psalmo . 78.... Inn disem psalmenn werdenn begriffenn Zwaierley fleisch so die mennschen honndt das naturlich vnnd das himlisch..."
20. For a discussion of Paracelsus' anthropology, see Daniel, "Invisible Wombs."
21. Psalm 79:2–4.
22. Psalm 79:11–13.
23. See his exegesis of Psalm 80 in particular: Paracelsus, *Ex Psalterio*, fols. 208b–209a: "Ex Psalmo 79. Qui Regis Israel intende. Inn disem psalmen begreifft der prophet Assaph das nachtmall Christi wie das selbig ann vnns lanngenn wirdt alß dann beschehen...Deus Virtutem conuerterer Aber hie mellt der phrophet die Ausserwelten vnnd redt onn derselbigen stat alß wolt er sprechenn aber Inn deiner tugennt vnnd der barmhertzigkeit laß vnns dein erweltenn darzu nitt kommen das wir der massenn werdenn wie die willtenn sew vnnd das felt fych sunder behalt vnns Inn deiner hanndt vor der selbigenn vppigenn leer vnnd suchents taglichen heim wie wir auß dem weinstockh wachsenn vnnd laß disenn zweig wachsen vnnd laß disenn zweig wachsen vnnd laß disenn wienstockh denn du selbst gepflanntzt."

24. See Dane Thor Daniel, "Paracelsus on the Lord's Supper: *Coena Dominj nostrj Jhesus Christj Declaratio.* A Transcription of the Leiden Codex Voss. Chym. Fol. 24, f. 12^{r}–29^{v1}," *Nova Acta Paracelsica* N.F. 16 (2002): 141–162.
25. For a summary of the subject of Paracelsus' Eucharistic thought (immortal philosophy), see also Daniel, "Paracelsus' *Declaratio* on the Lord's Supper," 141–162.
26. John 6:27. The Vulgate version reads: "operamini non cibum qui perit sed qui permanet in vitam aeternam quem Filius hominis vobis dabit hunc enim Pater signavit Deus."
27. Matthew 26:26. Vulgate: "Cenantibus autem eis accepit Iesus panem et benedixit ac fregit deditque discipulis suis et ait accipite et comedite hoc est corpus meum." This is repeated in Mark 14:22, Vulgate: "et manducantibus illis accepit Iesus panem et benedicens fregit et dedit eis et ait sumite hoc est corpus meum."
28. Daniel, "Paracelsus' *Declaratio* on the Lord's Supper," 146, n. 14, cited from *Coena Dominj*, fol. 16^{v}: "das brodt ist da ahnstatt deß leymen. dz ers prochen hatt, ist die vrsach, daß er nicht ein menschen machen will, sonder vil. darum dz brochen sein will einem yeglichen sein newe gepurt auß diesem brodt, dann Gott der Vater machet nur ein menschen."
29. See Katharina Biegger, "*De Invocatione Beatae Mariae Virginis.*" *Paracelsus und die Marienverehrung* (Stuttgart: Franz Steiner, 1990).
30. Michael Bunners' dissertation contains an excellent discussion of this topic: Bunners, *Die Abendmahlsschriften*, 94–96. Bunners writes, 94–95, "Sie ist unsterblich gen Himmel gefahren. 'Gott hat in das todtlich fleisch nit geheurat...Maria ist kein 'tödtlich (sterblich) Weib.'" He adds, 95, "Erst Weigel und Böhme haben, ihrer Denkart entsprechend, diese mariologischen Ausführungen des Paracelsus spekulativ weitergebildet." Andrew Weeks, *Paracelsus: Speculative Theory and the Crisis of the Early Reformation* (Albany: State University of New York Press, 1997), 83–85, evaluates Paracelsus' discussion of the four persons of the Godhead, a topic that Paracelsus abandoned after his very early writings.
31. Bunners, *Die Abendmahlsschriften,* 95. At the same time, Paracelsus reconstructed the biblical genealogy so that Abraham too escapes the tainted Adamic seed: Abraham took on a new seed, fashioned out of the new 'limbus.'"
32. Alexander von Suchten, *Chymische Schrifften* (Franckfurt am Mayn: In Verlegung Georg Wolffs, 1680), 305–356.
33. Toxites, "Vorrede," in Paracelsus, *Astronomia Magna,* 11–13.
34. See Allen G. Debus, *The English Paracelsians* (New York: Franklin Watts, 1966), 24, and Michael T. Walton, "Genesis and Chemistry in the Sixteenth Century," in *Reading the Book of Nature*, 1–14; 4.
35. Walton, "Genesis and Chemistry," 4.
36. Ibid., 7.
37. Debus, *English Paracelsians.* This is taken from a reproduction of part of Bostocke's *Difference between the Auncient hisicke... and the Latter Phisicke*

(London: 1585) in Allen G. Debus, "An Elizabethan History of Medical Chemistry," *Annals of Science* 18 (1962): 5–7.

38. Ibid., 8. Debus draws from the English translation of the preface to *Basilica chemica*: *The Admonitory Preface of Oswald Crollie*, trans. H. Pinnel (London: 1657), 22, 58.
39. Walton, "Genesis and Chemistry in the Sixteenth Century," 4. One should note several other pertinent studies, such as Siegrid Wollgast, "Zur Wirkungsgeschichte des Paracelsus im 16. und 17. Jahrhundert," in Peter Dilg and Hartmut Rudolph, eds. *Resultate und Desiderate der Paracelsus-Forschung* (Stuttgart: Franz Steiner Verlag), 113–145. See also Rudolf Schlögl, "Ansätze zu einer Sozialgeschichte des Paracelsismus im 17. und 18. Jahrhundert," in *Resultate und Desiderate der Paracelsus-Forschung*, 145–162; Jole Shackelford, "Seeds with a Mechanical Purpose: Severinus' Semina and Seventeenth-Century Matter Theory," in *Reading the Book of Nature*, 15–44; and Allen G. Debus, "Iatrochemistry and the Chemical Revolution," in Z.R.W.M. von Martels, ed. *Alchemy Revisited: Proceedings of the International Conference on the History of Alchemy at the University of Groningen, 17–19 April 1989*, (Leiden, New York, København, Köln: E.J. Brill, 1990), 51–66, esp. 58.
40. I, 12:102. "was nun mit der kunst [*membra*] geschehen muß, dasselbig muß durch den menschen volent werden. also muß magica scientia humana sein, auch nigromatia und die andern alle. nun merkent aber weiter, das noch zehen membra seind in der astronomia. es seind aber nicht artes humanae, sonder artes aethereae. das ist, der himel ist selbs astonomus, ist selbs derselbig, der es tut, und volbringt und macht das werk on menschliche hülf."
41. I, 12:16. Hence, Paracelsus has introduced an important qualifier, that this book does not treat the whole of cosmology, for astronomy does not deal with elemental theory. And yet, the elemental body *is* treated insofar as the firmament incarnates it spiritually. The discussion here is indeed about the effects of the heavens, that is, what pertains to the "aetherial spirits" and their actions. The rulers—the predestined parts—originate in the heavens, hence the governance by the stars, the subject of natural astronomy. Thus, Paracelsus will be able, for any object or being, to speak in terms of the interactions between elemental body and sidereal body, the former referring to the tangible corpus, the latter to the invisible-subtle-directing spirits.
42. I, 12:85.
43. Paracelsus, *Labyrinthus medicorum errantium*, in I, 9:208; 82. See also Kurt Goldammer, *Der Göttliche Magier und die Magierin Natur: Religion, Naturmagie und die Anfänge der Naturwissenschaft vom Spätmittelalter bis zur Renaissance, mit Beträgen zum Magie-Verständnis des Paracelsus* (Stuttgart: Franz Steiner Verlag, 1991), 46–47.
44. See Lynn Thorndike, *A History of Magic and Experimental Science*, vol. 6 (New York: Columbia University Press, 1941), 598: Gacaeus, *Astrologiae methodus* (Basel: 1576) lists Trithemius (b. 1462), who might have been

one of Paracelsus' teachers, as a "nigromanticus"; Caspar Busch Sr. is listed as a "magnus nigromanticus."

45. I, 12:91.
46. Ibid., 89–91.
47. Ibid., 92: "und die kunst signata lernet die rechten namen geben einem ieglichen, wie im angeboren ist."
48. Ibid., 97. "darumb von nöten ist, das die firmamentisch arznei gleich sowol als die irdisch erkent werde und gebraucht. dan zu gleicher weis wie aus der erden die arznei entspringen und wachsen, also auch as den sternen. wiewol ein solche underscheit hie gemerkt wird, das die arznei der elementen mit sampt einem corpus wechst, also das kein arznei aus den elementen wird, das die arznei der elementen mit sampt einem corpus wechst, also das kein arznei aus den elementen wird, alein sie hab dan ein corpus, in welchem die kraft ligt. aber die arznei des firmaments ist nit also, sie ist ein arznei on ein corpus, alein für sich selbs wol separirt und purgirt, bedarf keins alchemisten, der da separire purum ab impuro oder extrhire quintum esse von den vier elementen, sonder es ist ein arznei, zu dero nichts gehört, weder zu addiren noch zu decoquirn, als alein zu administrirn und dem kranken zu applicirn. der nun die krankheiten, so vom himel wachsen und werden, erkent und weiß in leib und wuntarzneien und durch die firmamentische arzneien denselbigen kan widerstehen, der heißt medicus adeptus. das ist sagax medicina, das ist der grunt, der die halb arznei regirt. dan dermaßen ist der himel mit dem elementischen leib vereinigt, das die infection von dem unsichtbarn in das sichtbare felt, und aber das sichtbar mag in das unsichtbar nicht wirken."
49. Ibid., 195. "Also wird nun probirt, das philosophia adepta von den menschen nicht mag gelernet werden, alein wol der buchstab von ime genomen, wie die kreuter von der erden haben. nun ist von nöten das wir wissen, was der philosophus adeptus sei, damit wir von im lernen. so wissent, er ist ungreiflich, unsichtbar, unentpfintlich und ist bei uns und wonet bei uns in aller gestalt wie Christus spricht: ich bin bei euch bis zu end der welt. und aber niemants sicht in, niemants greift in, noch ist er bei uns; also ist auch der philosophus adeptus bei uns. und wie Christus sagt: das reich gottes ist under euch und aber niemants sichts, also ist auch philosophia adepta under uns und niemant sicht sie. und wie Christus sagt: lernet von mir, also müssen wir auch lernen von dem philosopho adepto. und wie Christus sagt: es wird ein ieglicher schreiber von got geleret, also wird auch ein ietlicher naturalis vom philosopho adepto gelernet."
50. Ibid., 85.
51. Ibid., 158–159. "also ist kein ding nicht, es habe einen solchen geist, wie ich iezo erzelt hab, das ist kein corpus ist nit one einen schatten oder stime. und wie die sone einen ietlichen schatten machet, also ist ein andere sonne, die macht den schatten in der nectromantia und heißet sol gaba nale. nun kompt der selbig geist vom firmament und wird in der geburt naturlich angeborn und durch solche geist sen wir suchen. zu gleicher weis als einer, der einer stime nachgehet oder wie ein jeger einer spor nachgeht, die selbig

spor ist auch der geist, dan bei solchen dingen findet man und das suchen oder gespor nemen in mancherlei weg. nun seind solcher geist mancherlei art, ein ander dem menschen, ein ander dem ihe, ein ander den vögeln, ein ander den würmen, ein ander den metallen, ein ander den gemmis, ein ander dem holz, ein ander dem erz, ein ander den kreutern. welcher nun wil ein nectromanticus sein, der sol und muß wissen solcher geister, dan one sie wird er nichts finden. The spirits come out of the body and go to the sought item; they are magnetically pulled to what a person is seeking."

52. I, 12:142–148.
53. Ibid., 414.
54. Ibid.
55. Ibid., 429.
56. Ibid., 430.
57. Ibid., 432–433. "so wir wissent, aus was samen der donner wird, item der regen, item der hagel, item anderst alles, so mögen wir aus dem selbigen wol wissen und erkenen, aus was ursachen der teufel und seine geister solche wetter machen mögent. Dan das ist die meteorische philosophei, das aus dreien dingen also firmamentische impressiones geborn werden, als nemlich aus salz, aus sulphure, aus mercurio, die drei nach dem und sie die natur componirt, demnach wird das selbige generatum, als dan archeus firmamenti schmit und bereitet. so nun die drei ding der anfang himlischer impression seind, und das corpus und die materia, so ist es gleich zu verstehen, als einer der ein samen nimpt, und tregt in an ein andern ort, seet in, pflanzet in, tut im das, so im zugehört, so wird das daraus, das daraus werden sol. nun iezt folget aus dem, das zu gleicher weis wie wir menschen auf erden mit dem irdischen handeln, also die geist im firmament auch handeln mögen, also das sie mögen die simplicia im firmament nemen und die selbigen componirn zu einem donner, hagel, wetter, wolkenbruch etc. dan inen ist solches gleich so gewaltig, als einem steinmezen der einen turn bauet... also machen die bösen geist dergleichen wetter, stral, reif und anders mer... und aber als dan so wissent, das got solchs nit alzeit verhengt, sonder hindert. dan wo solchs nicht in dem gewalt gottes stünde, so würden wir nicht mügen auf erden bleiben. iedoch aber, so geschichts nicht on ursach, wo es sich also begibt, das solche teuflische wetter komen, es sei zusampt den schaden, den sie tun, ein praesagium und iudicium eines schentlichen, zukünftigen unfals, der dan nicht besser zuverstehn ist, als die seind, von denen solche wetter entspringen, gemacht und gefertigt werden.... aber in andern wege ist noch nicht erfaren worden, das der teufel und seine geist anderst haben mögen handeln in solchen dingen, oder inen zugeben, als alein in anzeigung zukünftiges dings, das so unfletig sein wird, als der ist, von dem es kompt. darumb ist es groß von nöten, das die werk, so von der hant gottes komen, wol verstanden werden."
58. I, 12:417.
59. Ibid., 418–419.

60. Ibid., 421. "er hat verstoßen die mechtigen vom stul (das ist, den Lucifer als den mechtigsten engel) und hat erhöhet die demütigen, das ist, die armen engel in der demut, hat er an der andern stat gesezt. dan das beweist sich hie auf erden das reichtumb wider got ist; . . . "
61. Ibid., 436. "wo aber der mensch im weg gottes nicht wandelt, da ist des selbigen geist beim selbigen ein regirer, also das der mensch wird, wie sein geist. also werden die besessenen, das ist, die flaschen gelerten, die betriegerischen handirer, also etliche gar von sinnen komen, als sich dan offenbar beweist."
62. See n. 10 above.
63. Karl Sudhoff, *Kritik der Echtheit der Paracelsischen Schriften*, I. Theil, *Die unter Hohenheim's Namen erschienenen Druckschriften* (Berlin: Georg Reimer, 1894), 647–648. See also 465 for Sudhoff's discussion of the first edition of this work: "Ausser den Capitelüberschriften und einigen kurzen Ueberschriften besonderer Abschnitte und einer kurzen Einleitung zum 4. u. 39. Capitel besteht das ganze Buch nur aus Stellen des alten und neuen Testamentes; kein Wort ist aus Hohenheim's zahlreichen theologischen Handschriften entnommen, auch findet sich nirgends eine handschriftliche Spur eines derartigen Werkes von Paracelsus. Die Unterschiebung ist zweifellos; . . . "
64. Ibid., 158.
65. Ibid., 160.

CHAPTER 2

Building Blocks: Imagination, Knowledge, and Passion in Agrippa von Nettesheim's *De Oculta Philosophia Libri Tres*

Allison B. Kavey[1]

De Occulta Philosophia Libri Tres proposes a world structure that is derived from, but certainly not identical to, the ones put forth by Neoplatonic natural philosophers and Catholic theologians. It intertwines these two intellectual traditions, smoothing over specific points of difference, in order to produce a malleable world that invites magicians to exercise their potential to manipulate it. These intellectual traditions contain two fundamental pieces: the structure of the world itself and the origin and flow of the power to induce natural change. By putting these puzzle pieces through some rigorous calisthenics, Agrippa conceives of a world imagined and willed by God but governed at multiple levels by different forces, including God himself, angels, demons, the stars and planets, and systems of sympathy. The power for individual natural philosophers, however, lies in the fact that nearly all of these systems can be influenced by an educated and devout magus.

Agrippa designed his model of the world around magical change, editing existing intellectual traditions that imagined and then restricted magical potential to produce a world in which human-induced magical change was implicit in its creation and all of its functions. This intentionality makes Agrippa's cosmogony significantly different from those proposed by other Renaissance natural philosophers interested in magic, including Ficino and Pompanazzi, both of whom imagined worlds in which magic existed within a strict hierarchy of forces that magicians could manipulate but never fully command. Agrippa adopted the notion of a hierarchy of forces, but he made it possible for the most capable and studious magicians to ascend the ladder and, in the end, sit next to God. In short, Agrippa's world was made by God for magicians, his most highly favored sons.

Agrippa had an extraordinary effect on Renaissance and early modern natural philosophy. Some of that impact was due to the vast circulation of his magical opus, *De Occulta Philosophia Libri Tres*, which was available in three versions throughout the majority of the sixteenth and seventeenth centuries. The juvenile manuscript of 1509/1510, which Agrippa originally sent to his mentor, the magical abbot Trithemius, saw significant circulation among those interested in the revival of magical thought.[2] The incomplete 1531 printed edition, which contained only book one and the table of contents for books two and three, also saw relatively impressive print circulation.[3] The complete printed edition in 1533 went through multiple editions, with subsequent printings in Lyon in 1567 and an English edition printed in London in 1651.[4] The spurious fourth book of occult philosophy was also responsible for some of Agrippa's popularity, and it developed a print life of its own from 1655 through the middle of the nineteenth century.[5]

Richard Nauert took up the task of contextualizing Agrippa's thought in terms of the intellectual and religious changes occurring during his lifetime, and while he did not focus especially on *De Occulta Philosophia Libri Tres*, he provided exceptional background information on Agrippa's complicated employment history and intellectual contacts.[6] Frances Yates adopted Agrippa as part of her argument on the influence of the hermetic tradition on Renaissance natural philosophers, though her reading of *De Occulta Philosophia Libri Tres* was relatively cursory.[7] Paola Zambelli devoted her dissertation and several significant articles to Agrippa's work; she attempted to bridge the apparent disconnect between his magical opus, which occupied twenty years of his intellectual life in writing and revision, and his skeptical diatribe, *De Incertitudine et Vanitate Scientiarum* (1530), in which he excoriated the social and moral limitations of Renaissance religion and thought and apparently renounced magic.[8] Vittoria Perrone Compagni produced an exceptionally useful edited edition of the 1533 version of *De Occulta Philosophia Libri Tres*, in which she noted the additions between 1509, 1531, and the final version, traced the sources behind the text, and proposed the importance of examining the book's structure for understanding its epistemology. Her work is notable for its refusal to accept the traditional characterization of Agrippa's magical opus as an encyclopedia that was useful, at best, for its insight into circulating ideas about magic in the early to mid-sixteenth century. In her discussion of the juvenile manuscript, which is most susceptible to this attack, she contends,

> Unlike its predecessors, *De Occulta Philosophia* proposes a total reorganization of magic as an umbrella science which, by gathering under a single

> roof all the cognitive data collected in the various fields of scientific research, would guarantee the effectiveness of each branch of research and make explicit its potential for acting upon reality.[9]

Her introduction serves as a very strong beginning for a close textual analysis of *De Occulta Philosophia Libri Tres*, but given its limited length, it cannot provide complete insight into Agrippa's rich intertwining of structure and content. I take her lead in this chapter, contending that close textual analysis is the only way to understand Agrippa's distinct cosmogony and the place of magic in it.

Since Compagni's edition, Agrippa has continued to attract scholarly attention. Christopher Lehrich made Agrippa's work, especially *De Occulta Philosophia Libri Tres* and *De Vanitate* the focus of his largely theological monograph, in which he contended that the magical thought in *De Occulta Philosophia Libri Tres* was intended to illustrate the author's, and presumably the readers', search for gnosis and the power it promised as they better understood themselves and God.[10] His argument depends upon a linear reading of the text and does not account for the complicated relationships among the chapters and the books. Recently, Wouter Hanegraaff has offered a significant revision of Agrippa's use and understanding of hermetic thought. He usefully contends that the magic that Agrippa detailed in *De Occulta Philosophia Libri Tres* was both significantly influenced by the Corpus Hermeticum and intended to extend the power of existing magical thought to such an extent as to make it unrecognizable, or "greater than magic."[11] He traces the notable influence of Lazzarelli's *Crater Hermetis* on Agrippa's ideas about the promised results of the search for gnosis when undertaken by a skilled and devoted adept, notably the attainment of superhuman power.

I agree with his contention, but I suggest that the degree to which Agrippa extended human potential to influence the world in his text makes it far greater than the "compendium of Renaissance magic" or even the "mature religious perspective" that Hanegraaff describes. It is, in fact, the work of a man whose quest for knowledge exceeded the hermetic goal of locating knowledge through study of one's self and God and represents a unique combination of religious and hermetic ideas about the type and direction of change in the natural world.[12] In book three, for example, Agrippa signals his awareness of hermetic ideas about the search for gnosis and its close relationship to religion when he writes, "This religion is a continual contemplation of divine things, and by good works a uniting of oneself with God and the divine powers."[13] Agrippa's magical opus illustrates the tremendous advantages to be had from combining theology with natural philosophy, allowing adepts to not only better understand themselves and God but also to access some of the forces that He embedded in the world to govern it.

Both Christopher Lehrich and Wouter Hanegraaff classify *De Occulta Philosophia Libri Tres* as a magical encyclopedia whose real importance lies in its theology, despite the significant impact it had on later Renaissance and early modern magical thought. They are partially correct. I agree that Agrippa saw a great deal of room for magic to improve its standing in intellectual and religious circles, where it was frequently despised as cheap charlatanry or decried as the work of a heretic, but the depth of his project was greater than simply returning magic to its status as a respected discipline. I also freely acknowledge that Agrippa mined his material from a tremendous array of existing Neoplatonic, hermetic, and religious texts; he wrote absolutely nothing new about the creation of the world, the forces that govern it, or the major knowledge systems that describe it. He did, however, organize existing knowledge systems into a fundamentally new structure of the world that makes the claim that his magnum opus was an encyclopedia laughable. The significance of the text lies in the world it proposes, which depends upon an equal combination of theology and occult philosophy, and results in a structure in which the potential for human magical alteration was foreseen at the creation and embedded into the forces of change influencing the celestial and natural worlds.

The Structure of Agrippa's World and the Structure of His Book

The three definitions of magic that begin book one offer a preliminary structure for *De Occulta Philosophia Libri Tres*, which is itself divided into three books. The titles of the books are derived from Aristotle's division of speculative philosophy in the *Metaphysics*, and they correspond to the three types of magic.[14] The reader slowly ascends from natural magic through mathematical calculations and finally arrives at theological magic, where he will learn about and then master the fundamentals to command preternatural beings to accomplish his will. This superficial structuring device quickly falls apart, since Agrippa does not separate his definitions of magic as neatly as he promises at the beginning. Instead, all three books are dedicated to a recitation of the potential to be found in the crossings over between and among the three branches of magic.

He returns to this critical part of his project throughout book one, intentionally reminding readers of the rich interconnections linking the theories underpinning Neoplatonic and Judeo-Christian thought. While these intersections were far from unfamiliar to most Renaissance readers, the consistency with which they occur beside each other in multiple chapters of book one, which is, according to the Agrippa's definitions of magic, simply discussing natural philosophy, serves as a strong correction to any who imagined a natural world in which God's will and imagination did not continue to play an active role. In book two, he provides a series of Neoplatonically influenced and theologically informed ideas about the importance and utility of numerological, astrological, and symbolic magic, accomplishing the remarkable task of making religion out of magical acts. Book three begins with theology, but within a few chapters it has placed religion and magic back in the same arena, indicating that no magical act can be done without faith in divine power, that all faith is powerful, even faith in gods other than the true one, and that magical practice, when done by a worshipful adept, is its own form of worship.

Given the slipperiness of magical and theological categories in *De Occulta Philosophia Libri Tres*, a more useful structuring device for Agrippa's world than the three kinds of magic is clearly required. I propose one derived from the Trinity and shared with Neoplatonic thought about the creation of the world. The highest realm, which is assimilated to the Trinity but especially God the Father, is linked to the imagination and is the arena of forms or divine ideas. Agrippa notes the critical role played by divine imagination in the creation of the world, asserting in his chapter on the nature of the Trinity:

> Hence we read in Plutarch that the Gentiles described God to be an intellectual and fiery spirit, having no form, but transforming himself into whatsoever he pleases, equalizing himself to all things; and we read in Deuteronomy, our God is a consuming fire; of whom also Zoroaster says, all things were begot of fire alone; so also Heraclitus the Ephesian teaches; hence Divine Plato has placed God's habitation in fire, namely understanding the unspeakable splendor of God in himself, and love about himself;[15]

In book one, Agrippa introduces his readers to the creative power of fire, which he identifies as the source of all life and a critical tie uniting divine and human creative power and the divine, preternatural, and natural realms.

> The properties of the Fire that is above are heat, making all things Fruitful and light, giving life to all things. The properties of the infernal fire are a parching heat, consuming all things, and darkness, making all things barren. The Celestial and bright Fire drives away spirits of darkness; also this our Fire with wood drives away the same, in as much as it has an Analogy with, and is the vehicle of, that Superior light; as also of him, who said, "I am the Light of the World, which is true Fire, the Father of Lights, from whom every good thing that is given comes; sending forth the Light of his Fire, and communicating it first to the Sun and the rest of the Celestial bodies, and by these, as by mediating instruments, conveying that light into our Fire.[16]

Finally, he illustrates the links between God the father—the source of all creative imagination—and the Holy Spirit—the font of creative passion—in his discussion of the location, meaning, and utility of occult properties in the natural world.

> Hereunto may be added that in the Soul of the world there are as many Seminal forms of things as Ideas in the mind of God, by which forms she in the Heavens above the Stars frame to herself shapes also, and stamped upon all these same properties; on these Stars, therefore, shapes, and properties, all virtues of inferior species, as also their properties do depend; so that every Species has its Celestial shape, or figure, that is suitable to it, from which also proceeds a wonderful power of operating, which proper gift it receives from its own Idea from the seminal form of the Soul of the world.[17]

Agrippa's God the father is the source of all things in their most perfect form, as well as the original center of creation and destruction. He has the power to imagine the world into being and to tie all of his creations together through an elaborate web of sympathies uniting His imagination with the things he has imagined for all eternity.

The middle section of the Trinity, which obeys the direction of God the son, is driven by intelligence or knowledge and is the arena of magical, mathematical, and astrological action. It reflects the divine knowledge that created the world, but it also oversees the knowledge webs that emerge through human interaction with the natural world, and between preternatural and natural beings. God the son allows the world to work in a predictable, rational way that can be understood by those who study nature. Agrippa's ideas about God the son can also be found in his discussion of the Trinity.

> for the divine mind by itself with one only and uninterrupted act... generates himself and Issue and Son who is the Full Intelligence, complete image of himself, and the perfect pattern of the world. Also Mercurius Trismegistus in Asclepius mentions the Son of God in diverse places, for he says my God and Father begat a Mind, a work diverse from himself... and in Pimander (where he seems to prophecy of the Covenant of grace to come, and of the mystery of regeneration) says, the author of Regeneration is the Son of God.[18]

The knowledge of the world embodied by God the son is a perfect knowledge of all of the original creation and the systems of sympathy and antipathy tying together the celestial, preternatural, and natural realms. An understanding of God the son can, therefore, be obtained by studying the world and the causes of change, and thus the lowest rung of the ladder that every magus must climb toward the divine mind begins with the study of the thing it created, and then a comprehensive understanding of the knowledge behind it.

Numerology as an epistemological system illustrates the shift from the natural to the preternatural and the visible to the hidden, and in book two Agrippa outlines the divine knowledge that lies behind the immense virtues in numbers and the power to be obtained by a perfect understanding of them.

> For it is no wonder, seeing there are so many, and so great occult virtues in natural things, although of manifest operations, that there should be in numbers much greater, and more occult, and also more wonderful and efficacious, for as much as they are more formal, more perfect and natural in the celestials, not mixed with separate substances; and lastly, having the greatest and most simple commixtion with the Ideas in the mind of God, from which they receive their proper and most efficacious virtues: wherefore also they are of most force, and most conducive to the obtaining of spiritual and divine gifts, as in natural things, elementary qualities are powerful in the transmuting of any elementary thing.[19]

God the son provides an entrée into the Trinity for those magi sufficiently devoted to their quest to look beyond the natural world to the causes governing change within it and the very origins of change itself.

The lowest of the spheres also reflects the ideas of the Holy Spirit and is the playground of divine creative imagination, in which all of the forms were originally conceived. This creative energy perpetuates life on earth, driving combinations and exchanges that produce new things as well as allowing existing things to continue. It provides a future for God the father's original creation, and it produces the canvas on which God the son's knowledge is displayed.

> From this complete intelligence of supreme fecundity his love is produced, binding the intelligence with the mind. And by so much the more, by how much it is infinitely more intimate to itself than offspring to their parents. This is the third person, viz. the holy spirit.[20]

He also contends that the spirit unites God the father with God the son and their created worlds, allowing for a unity through trinity and the endless potential for creation through the interactions of the three parts of the Godhead.

> Therefore it is God, as Paul says, from whom, in whom, and by whom are all things: for from the Father as from a fountain flow all things, but in the Son as in a pool all things are placed in their Ideas, and by the Holy Ghost are all things manifested, and everything distributed to their proper place.[21]

The three books of *De Occulta Philosophia Libri Tres* correspond to these divisions more neatly than they do the three types of magic, though even here the connections between books and realms reflect more commonalities than differences. This suggests that a world composed of linked spirals, rather than a hierarchy, is a more useful spatial layout for Agrippa's cosmogony. It corresponds more fully to the multidirectional change he highlights throughout *De Occulta Philosophia Libri Tres* and the significant areas of commonalities among the spheres, as well as illuminating the critical areas of engagement he envisions between the theories underpinning theology and Neoplatonism.

The Forms Beneath the Structure

Two versions of the world frame Agrippa's cosmogony; those proposed by Plato and developed by Neoplatonic natural philosophers and those conceived of in the Old and New Testaments and elaborated upon by Christian and Jewish theologians. These two models appear consistently throughout *De Occulta Philosophia Libri Tres*, lying beside each other in chapters devoted to everything from the influence of the heavens on the earth to the power of specific numbers. Their incessant coupling elides their often significant differences and supports Agrippa's model of the world, in which these intellectual systems work in tandem, rather than in competition. Agrippa contends that human knowledge systems, especially those developed by Plato and the Hermetic philosophers, provided magicians with the potential to access the divine knowledge structures outlined in the biblical texts. Divine gnosis created the world, along with divine imagination and passion, but human gnosis developed to allow the most erudite students the chance to recreate their worlds using their own imagination and passion.

The first coappearance of religious and natural philosophical knowledge systems comes in book one, chapter one, which describes the structure of the world.

> Seeing there is a three-fold World, Elementary, Celestial, and Intellectual, and every inferior is governed by its superior, and receives the influences of the virtues thereof, so that the very original and chief worker of all things does by Angels, the Heavens, Stars, Elements, Animals, Plants, and Stones convey from himself the virtues of his omnipotency upon us.[22]

The threefold world is a staple of both Neoplatonic and Judeo-Christian thought, and each intellectual tradition imagined it to its own advantage, with the first conceiving of change going from the top to the bottom through richly developed systems of sympathy, and the second proposing that God initiates all change and sends it down through his various messengers, beginning with the highest orders of angels and continuing through the stars to the world below. Agrippa's first revolutionary proposal for these models comes in the very next sentence of chapter one, in which he contends that change can be initiated from any point in the threefold world, rather than only at the top.

> Wise men have conceived it in no way irrational that it should be possible for us to ascend by the same degrees through each World, to the original world itself, the Maker of all things, and first cause, from whence all things are and proceed, and also to enjoy not only the virtues, which are already in the more excellent kinds of things, but also beside these, to draw new virtues from above.[23]

This sentence manages to undermine every assumption in the two contributing models of the world about the direction of natural change, the potential for magic, and the defining role of magicians in creating and directing change. It asserts that the threefold model does not depend upon the one dimensional relationship of superiors governing inferiors, but in fact contains a much richer system of sympathy that permits change to be introduced at any point and to be felt at all levels. Furthermore, it proposes that the systems of sympathy should be studied at the lowest level of the world (the elementary or natural) in order to ascend to the next level (the celestial), and finally to the divine realm.[24] This excerpt combines Old and New Testament thought about the creation with Neoplatonic philosophy to imagine a world in which magical potential was imagined and rewarded at the very instant of creation, so that students of nature could, if sufficiently disciplined and devoted, rise through the levels of the world to attain the status of God, and draw down new divine virtues, recreating the world in the process. The rest of *De Occulta Philosophia Libri Tres* strives to correct that mistake by exploring at great length and in intricate detail how the world was conceived and constructed specifically to permit the greatest possible room for pious and learned magicians to exercise their craft.

God is the obvious place to start, as he created the world and designed the forces that govern it and the systems through which they operate. Agrippa conceives of no difference between theology and magic, arguing that both represent epistemological systems that provide insight into God. In this erasure lies the key to understanding Agrippa's model of the world, which rests upon the belief that God imagined magic and designed the world to accommodate it, rather than being offended by it and its practitioners. His cosmogony was remarkable for exactly this reason.

The drumbeat uniting theology and natural philosophy begins in book one, chapter two, which defines magic according to the three academic disciplines a magus must master to practice it; natural philosophy, mathematical philosophy, and theology. The first "teaches the nature of those things which are in the world, searching and enquiring into their Causes, Effects, Times, Places, Fashions, Events, their Whole and Parts."[25] This is a conservative definition of the discipline, limiting its practitioners to the study of the world they can observe and experience, and limiting their arena of inquiry to the world they inhabit. The only potential that Agrippa perceives for those who master this most limited form of philosophy is the apprehension of the systems governing change in nature, which seems limited except in the face of book one, chapter one, in which every relationship between two objects becomes a step on a ladder into the mind of God.

Mathematical philosophy "teaches us to know the quantity of natural bodies, as extended into three dimensions, as also to conceive of the motion and course of celestial bodies."[26] This discipline reflects the combination of mathematics and astrology that reflected the middle level of intellectual accomplishment in the university period in this period. It contends that celestial bodies move in comprehensible and predictable patterns that can be studied and linked to observable events in the natural world. Again, he proposes unidirectional change, in which the heavens influence the earth. The only opportunity to undermine this argument comes in his suggestion of three-dimensional movement, which allows for change in a multiplicity of directions. Finally, he arrives at magic and theology.

> Now theological philosophy, or divinity, teaches what God is, what the Mind, what an Intelligence, what an Angel, what a Devil, what the Soul, what Religion, what sacred Institutions, Rites, Temples, Observations, and sacred Mysteries are. It instructs us also concerning Faith, Miracles, the virtues of Words and Figures, the secret operations and mysteries of Seals, and as Apulius says, it teaches us rightly to understand and be skilled in the ceremonial laws, the equity of holy things, and the rule of religions.[27]

The traditional study of theology would certainly accomplish everything listed in the first three lines, but by the fourth, Agrippa has blurred the line between religion and magic and drawn his readers into an exceptionally powerful field, in which magical arts can be accomplished with divine approval and power.

Math before Magic

The structure of *DOP* bullies its readers, drawing them into what seems like a logical and frequently familiar discussion of hermetic principles and then flinging them out every few chapters with a radical shift in subject matter or approach. This is not just true within each book; it is especially apparent to anyone who attempts to read straight through the entire volume. Books one and three, as I have demonstrated above and will elaborate upon below, largely fit together. They share subject matter, namely the nature of the world, the forces governing it, and the role of God and magic in it, and they are linked by vocabulary and approach. They also talk to each other, with chapters within both books referring back and forth to topics and questions raised in the other. This is no accident: Agrippa intentionally united two epistemologies that were too frequently understood to be separate, or at least linked through dependence, Catholic theology and Neoplatonic hermeticism. He deliberately schooled his readers in a worldview that included both knowledge systems, and he invited them to study both philosophies in order to understand the world, and their place and potential in it. But what can be said about book two, which promises action and gives instead a strange mix of theory and practice? Anyone who managed to plow through the apparently careless accumulation of Neoplatonic thought in book one might, perhaps, heave a great sigh of relief upon arriving at book two. Here, at last, is something practical, a real "how-to" guide to the manipulation of nature.

The book invites readers in with the promise that they will find "the mysteries of the Celestial magic, and all things being opened and manifested."[28] Instead of a clear delineation of the power of numbers over the natural world and how to use it, however, the book offers a lecture on why numbers are powerful and how they are linked to original divine intention. Despite the lovely scales, which go on forever, and the magical seals that conclude the chapter, almost no practice is revealed in these pages. In fact, no one who reads only book two would be able to perform magic or learn how magic fit into the world. Book two both promises and refuses to stand alone because it is the bridge between the two epistemologies introduced in book one and united in book three. It is the practical application of divine knowledge to the natural world, illustrated at great length not so readers can take advantage of it and do it themselves, but so they can appreciate what God has already done. As a result, book two flirts with cabala, the knowledge system that most clearly demonstrates the presence of and potential for God's presence in the world, but it leaves the topic unfinished, providing a clear bridge to book three, where it is developed in much more detail. Instead, it provides a basic education in numerology and astrology, including a collection of Neoplatonic and theological theories about the meaning and significance of particular numbers, their relationship to the Trinity and specific planets, and how their forces could be channeled through the uses of signs and seals. His assertion of the similarities between natural philosophy and theology in the first chapter of book two reminds readers that the structure of the world they want to change was designed by God and can be comprehended by studying mathematics, which contains the systems governing nature.

> But amongst all Mathematical things, numbers, as they have more of form in them, so also are more efficacious, to which not only Heathen

> philosophers, but also Hebrew and Christian divines, do attribute virtue and efficacy, as well to effect what is good, as what is bad.[29]

His claim that numbers "have more of form in them" reflects his Platonic roots, in which mathematical symbols have more in common with the forms in the Holy Spirit's imagination than the illusory impressions gained through observation. He reaffirms this in the second chapter of book two, in which he identifies numbers as "having the greatest and most simple commixtion with the Ideas in the mind of God, from which they receive their proper and most efficacious virtues."[30] He continues through the first fifteen chapters of book two to trace the occult powers, derived from divine contact, hermetic reference, or philosophical insight, contained in many numbers, including one through sixteen, eighteen, twenty, twenty-two, thirty-two, forty, sixty, seventy two, one hundred forty four, and one thousand seven hundred twenty eight. The combination of natural philosophy and theology is evident in all of these descriptions, but the number four will illustrate the point.

> The Pythagoreans call the number four Tetractis, and prefer it before all the virtues of Numbers, because it is the foundation, and root of all other numbers, whence also all foundations, as well in artificial things as in natural, and divine are four square, as we shall show afterwards, and it signifies solidity, which also is demonstrated by a four square figure. For the number four is the first four square plain, which consists of two properties, where the first is of one to two, the latter of two to four, and it proceeds by a double procession and proportion, viz, of one to one and two to two, beginning at a unity, and ending at a quaternity, which proportions differ in this, that according to arithmetic, they are unequal one to the other, but according to geometry are equal. Therefore a four square is ascribed to God the Father and also contains the mystery of the whole Trinity.[31]

The scale that follows this explanation details the importance of four in the natural and preternatural worlds, including the key parts of plants (leaf, stem, flowers, root), four metals (gold, silver, mercury, and lead), types of stones, elements, humors, judiciary powers, moral virtues, senses, elements of spirit, demonic and angelic princes, and rivers in hell. Both the description and the scale of four illustrate the centrality of natural philosophy and theology to Agrippa's model of the world, which resonates as much with ancient Greek ideas about the formation of the world according to harmonic principles as to the meaning and effect of the Trinity.

This combination, which seems awkward to modern minds and has bedeviled modern scholars of this text, is in fact logical given the intellectual world Agrippa occupied. He saw no contradiction between natural philosophy and theology and was frustrated when his Church and some of his peers chose one over the other. The underlying principle of *De Occulta Philosophia Libri Tres* is the importance of reconciling these two philosophies about the creation and structure of the world in order to allow magicians the greatest access to the occult forces they would need to access to create radical natural change. The book devotes a tremendous amount of print to making readers understand that the acts of learned magicians, like the rites of religious authorities, are in fact means of paying tribute to and accessing divine power. The arts of astrology and numerology illustrate the significance of these similarities, but cabala, which integrates them into a single, poorly understood, knowledge system designed to worship and assimilate God's strength is the ultimate representation of the power inherent in combining two such rich epistemologies.

Book two accomplishes exactly what Agrippa designed it to do: it invites readers in to view the presence of God in the world, then smacks them very hard for even thinking of using that knowledge to benefit themselves. Magic is not for the greedy, the faint of heart, the slow-witted, or the easily bored. It requires a patient student who loves his books more than himself and is willing to submerge his selfish desires in an exceptionally taxing quest to understand and finally enter the original forces that created the world. The structure of *De Occulta Philosophia Libri Tres* intentionally manipulates its readers into abandoning their quest by skipping ahead for the "good parts" and missing the key lines embedded one at a time in nonconsecutive chapters. This approach was well-loved by occult philosophers, alchemists especially, who wanted to test their readers' resolve and make sure that the uninitiated and unworthy would not gain powerful knowledge that they did not deserve and would abuse.[32] Agrippa borrowed the idea from any number of the books he mined for his great enterprise, but he did something incredible with it—he seeded his easily dismissed "encyclopedia" with the germs of a new world and offered his most disciplined readers the chance to play with it using the forces that created it.

Magic: Creation and Recreation

The study of word and symbolic magic was a significant preoccupation of late medieval and early Renaissance magicians, who devoted themselves to understanding the various operations of magical words and seals and the magical reasons for their effects.[33] This field would have occupied only a minute role in any traditional theology curriculum, if it had been entertained at all, in relation to the power of the names of the Father, the Son, and the Holy Ghost to drive away evil spirits or purify an area. In traditional Judaism, however, the magical virtues of words and figures occupied a much larger intellectual space because of the discipline of cabala, which was slowly becoming an important component of early Renaissance

Neoplatonism. Within the bounds of this small and poorly understood discipline, which combines numerology, word magic, and an extraordinarily detailed comprehension of the Old Testament, can be found the kernel of Agrippa's conception of a combined magical and theological cosmogony. God made the world and embedded within it pieces of his mind, which can be accessed by students of his word and create natural change.

Only in book three, where the discussion of the Trinity, the interconnections between faith, the "true faith," and God's role in governing the world continues, does cabala reach its full potential. Agrippa recounts that a great deal of power can be activated through the manipulation of the letters of his Hebrew name, which command everything that was created in its honor. The Hebrew characters have individual power as symbols, but that power is derived from the fact that they, in concert, produce the symbol for the divine name of the creator. Furthermore, that power was present in the symbols and belief systems of the ancient Greeks, as well as the Hebrew cabalists and the Egyptian magicians.

> Whosoever understands truly the Hymns of Orpheus and the old Magicians shall find that they differ not from cabalistical secrets and orthodox traditions; for whom Orpheus calls Curets or unpolluted gods, Dionysus names Powers; the Cabalists appropriate them to the numeration Pahad, that is to the Divine fear: so that which is Ensoph in the Cabala, Orpheus calls Might, and Typhon is the same with Orpheus, as Zamael in the Cabala.[34]

Christianity added to the range of name-based divine magic by making the symbols for the names of the other members of the Trinity as powerful as those used to spell out the name of God the Father. Thus, the names of God, the Son, and the Holy Spirit have independent magical power that can be accessed by magicians striving to change the sections of the world for which these two members of the Trinity are responsible. The names of the major angels and the Old Testament fathers also have significant magical potential when symbolically depicted and used by a magician with great skill and piety.

> And these are the hidden secrets concerning which it is most difficult to judge, and to deliver a perfect science; neither can they be understood or taught in any language except Hebrew, but seeing the names of God (as Plato says in Cratylus) are highly esteemed of the Barbarians, who had them from God...for they are the mysteries and conveyances of God's omnipotency, not from men, nor yet from angels, but instituted and firmly established by the most high God...and breathe forth the harmony of the Godhead, being consecrated by the Divine assistance: therefore the creatures above fear them, those below tremble at them, the

> Angels reverence, the devils are frightened, every creature doth honor, and every Religion adore them; the religious observation and devout invocation with fear and trembling yields us great virtue, and even deifies the union, and gives a power to work wonderful things over nature.[35]

The magician thus becomes a master of theology, uniting all of the faiths that have existed in combination with attempts to manipulate nature through the use of God's name. Magic becomes quite literally the study and worship of God, because it is impossible to study the world without studying the Trinity.

The design inherent in the creation is the focus of several chapters in book three, which concentrate on the original intentions behind the structure of the world and the systems of influence that continue to carry them out. It reads a divinely inspired world through a Neoplatonic lens, allowing for the combination of the threefold world defined intellectually and a threefold world divided according to its creation and the Trinity. It demonstrates the possibility inherent in both intellectual systems, indicating that neither eliminates the other, but rather that the two together are stronger, more functional, and offer more explanatory power about the nature of the world than either one alone. One of the most fundamental aspects of this cosmogony lies in the flow of power from God to other agents, which does not simply include God's will filtering through the layers of the threefold world but also involves the generation of independent ideas by other supernatural and preternatural entities. *De Occulta Philosophia Libri Tres*'s cosmogony relies upon a multiplicity of divine powers, all of whom command particular arenas of the natural world and require appropriate worship by magicians hoping to borrow their influence.

> It is necessary therefore that Magicians know that very God which is the first cause and Creator of all things, and also the other gods or divine powers (which we call the second causes) and not to be ignorant, with what adoration, reverence, holy rites, conformable to the condition of every one, they are to be worshiped.[36]

The claim for multiple gods allows Agrippa to resurrect the religious systems proposed by the ancients, especially the Egyptians and the Greeks, and to reconcile them with Catholicism by arguing that the multiplicity of gods worshiped by ancient pagans was present in the Church in the form of angels, the highest tier of which (the Intelligences) were so closely tied to God as to be inseparable from Him. He also locates the Trinity in the works of ancient philosophers, most notably the Platonists and Plotinus, and less surprisingly in the writings of Hermes Trismegistus and the Pimander.

The recreation of the entire world, however, is rarely the subject of a magicians' work. Most magic aims to influence natural objects or relationships by directing the forces, whether natural or supernatural, that govern them. The worship of divine powers is obviously required to accomplish even the simplest magical task, but their co-option can be overlooked in favor of borrowing some power from lesser entities. Agrippa's cosmogony permits this as well, since it details the flow of power from the Trinity, especially God, the father and God, the son through a series of angelic realms, the planets, and finally to the lesser angels and the demonic powers.

> Therefore the heavens receive from the Angels that which they dart down, but the Angels from the greater name of God and Jesus, the virtue whereof is first in God, afterward diffused into these twelve and seven Angels, by whom it is extended into the twelve signs, and into the seven planets, and consequently into all the other Ministers and instruments of God.[37]

Three angelic realms form a link between God and the planets, in the case of the first two, and between the planets and natural objects, in the final instance. The highest legion of angels is the supercelestial, the closest to God; these angels have no bodies and consist only of minds that direct God's will to the heavens and the lower angels. They are so powerful and so deeply embedded in divine intention that Agrippa concedes they can be considered gods themselves. The next angelic rank consists of the celestial intelligences or worldly angels, which take divine will as interpreted by the supercelestial intelligences and distribute it directly to the heavens, which take it down through their systems of influence to the natural world. Magicians wishing to influence these angels must use the angel's name, seal, and character when doing their spells.[38] The very fact that these angels, who are so close to the divine intelligence, are accessible to magical importuning suggests the depth to which magical potential is embedded in Agrippa's cosmogony. He clearly understood the world to be composed around magical potential, and conceived that every level of the universe can be altered by magical manipulation. This differentiates his conception of magic from most of those circulating during the Renaissance, including Ficino's, which held that only preternatural beings, the lowest angels and demons, could be reached by magicians. Agrippa envisioned greater reach for magicians, extending their voices to reach God's highest messengers, and through them, divine passion, knowledge, and imagination.

Agrippa described the least powerful angels with considerable detail, noting that these beings are the closest to earthly happenings and thus are the nearest link to divine forces. The lowest rank of angels is called Ministers, and they have direct power over natural objects and thus act as intermediaries between the planets and the natural world. They are divided according to their nature into the four elements, with airy or aerial angels impacting rational thought and action, fiery angels impacting contemplative or passive things, watery angels influencing the imagination, and earthly angels governing passion. These angels are easily accessed by magical rites and incantations that generally include their name, seal, and character and can be used to alter the outcome of a battle, influence love, or guarantee or doom the success of a business venture.[39] The simple magic to which these angels are susceptible differentiates them from their more powerful brethren, who must be worshiped like gods to sway their favor, and suggests that magic itself has a hierarchy, with simple incantations and symbols holding power over the least powerful angels and demons, and much more complex regimens, which greatly resemble religion, required to command more powerful entities. The magical hierarchy mirrors the one Agrippa proposes for the relationship between God, his angels, and later, his demons: the Trinity contains the passion, knowledge, and imagination that created the world, the angels, from the highest to the lowest ranks, direct divine ideas through the heavens to the world, and the demons strive to distract and destroy human potential.

From Angels to Demons

The ranks of preternatural beings are not, of course, limited to angels, and Agrippa's willingness to include demons in his cosmogony and to frankly discuss their potential to influence the natural world got him a significant reputation as a master of black magic. He depends upon both theological and natural philosophical descriptions of demons for his discussion, arguing first, according to Plato, that they are "angels who failed, revengers of wickedness or ungodliness, according to the decree of divine justice."[40] This excerpt is odd, since it is difficult to tell whether he meant that the angels were wicked or ungodly and thus, having violated divine justice, failed, or whether these angels, for whatever reason, failed and became avengers of divine injustice. The latter interpretation supports the broader cosmogonical picture in the book, since Agrippa carefully fit every preternatural being into divine intentionality and would have been more likely to keep these powerful beings as active agents than to eject them entirely from his power. This interpretation is further enforced in book three, chapter eighteen, in which they were intended, at the creation, to tempt souls and torment the fallen.

Another interpretation depends upon demons' infamous willfulness, which caused them, according to the Bible, to leave their place by God's side because they wanted independent worship and the potential to wreak their own kinds of havoc. Their malevolent independence supports this idea, since it is their very wickedness that makes them so inclined to engage with human affairs. These capricious and inherently nasty entities seem to have much the same power as the least powerful angels, bringing their malevolence to bear on natural objects, causing locusts to destroy crops, diseases to decimate populations, horses to rear and crush offending riders, battles to turn into bloody disasters, and desire to slide into adultery. Their influence can be occult, but it can also be easily apprehended by their victims.

The structure of the world in *De Occulta Philosophia Libri Tres* depends upon Agrippa's twin convictions that the original creation stemmed from such powerful forces that their echoes can still be found in every realm, and that only divinely inspired and directed magic will succeed in any meaningful way to change the world. The struggle between good and evil is recast through this lens, though the pieces that go into it are very familiar. Agrippa, like many Christians, was convinced that every person is caught between his governing angel and demon, who struggle to direct his soul, either toward God or toward the Devil. Both the angel and the demon are doing God's will, since the creation already contained the materials for both entities and the concept of human free will.[41] Magicians are as easily tempted as other people by the promise of tremendous power, eternal life, and piles of gold, and Agrippa cautions adepts to distinguish between the godly appreciation of divine work and occasional, humble imprecations for angelic intervention and the accumulation of individual power and might that results from frequent trafficking with demons. Though magic can be helped by both demons and angels, Agrippa asserted that only magicians who have attained the hermetic goal of uniting their souls with those of the Trinity will be able to recreate the natural world according to their own laws, while those who have been seduced by demons might have tremendous power over natural objects but will fail should they attempt to redirect the heavens. That failure comes from their own will—they will be more likely to settle for temporal gain, such as the accumulation of wealth and power, because the demons who command their souls will prevent them from seeking divine light and truth, which might save them. Their knowledge, which would necessarily have been hard-won and extensive, would have been wasted, distracted from uniting with imagination and passion, and then ascending toward the Trinity, by lowly human desires for power, wealth, and sex. Magic is not for the weak, the easily distracted, or the faint of heart.

Conclusion: "Mistress of His Own Intention"

The question of free will kept medieval theologians entertained for uncounted pages; why would an all-powerful God create sentient independent beings with the potential to dismiss and mock Him? Agrippa read those medieval theologians and he saw in their arguments the intent behind the original idea; free will exists in every particle of Agrippa's world, but it is not the will to act, but the potential to imagine, know, and conceive of brand new worlds. Without this, people would have no need for magic, since they would accept what God created and revel in the opportunity to enjoy divine knowledge, imagination, and passion. This happy paradise didn't even make it past the angels in Agrippa's cosmogony, since they also possessed imagination, knowledge, and passion, and while directing God's will through the heavens and onto the earth, they tweaked the world, and the Word, to reflect their own imaginations, knowledge, and passions. At least, they had the potential to do so, just like the people, the animals, the plants, and even the stones that received divine emanations through angelic emissaries and saw in them the chance to do something new, to create new worlds.

The story of Agrippa's cosmogony is not one of a single world, but of a stream of potential worlds all coexisting through the extraordinary art of magic, which brings earthly beings back into contact, and then into command, of the forces behind the original creation. It makes gods of men, offering beings with the will to imagine, the discipline to learn, and the passion to create the chance to make new worlds for themselves. It does so from familiar building blocks, and it tricks unwary readers, those from the twenty-first century as easily as those from the sixteenth, into believing that it offers nothing new to the overcrowded library of Renaissance magic. The structure matters as much as the content in *De Occulta Philosophia Tres* because the former gives new meaning to the latter, much of which was in circulation at the time that the text was written. This chapter has led you through the labyrinth to contend that Agrippa proposed a powerful union of theology and natural philosophy that empowered human potential to a far greater extent than any other book of its time; that the world built in *De Occulta Philosophia Libri Tres* has magic embedded in it at every level because the divine imagination wanted it that way; and that magicians are responsible for mastering their desires in favor of their passions, imagination, and knowledge, forsaking worldly riches and power for the chance to recreate the world.

De Occulta Philosophia Libri Tres has largely been dismissed by scholars as a clumsy and unreadable encyclopedia, attracting interest for the wide array of sources it consolidated into a single place rather than for its own sake. Read as a world-building text, however, it is outstanding in its originality and the potential it imagines for magicians to remake the world using their own versions of the original forces that created it. Unlike the utopias discussed by Giglioni in this book or the recast creation stories proposed by Coppola, *De Occulta Philosophia Libri Tres* combines a known story of the creation with a new twist on the forces God commanded to make it happen, and he envisions a world in the midst of a million potential worlds, rather than isolated from other forces or permanent in its current iteration. As a cosmogony, it is unique for the mutability of its proposed world, and for the tremendous array of forces that can alter it to suit their own visions. Its structure is also unique, combining intellectual traditions into a richly satisfying explanation of how and why the world works the way it does. Understanding and fitting ourselves into the world embedded in the structure of this text makes magicians of us all.

Notes

1. Thanks to the Chemical Heritage Foundation for the Roy G. Neville fellowship for the summer of 2005, which allowed me to begin working on *De Occulta Philosophia Libri Tres*. Many thanks also to Larry Principe and Vaughn Huckfeldt for their help in improving this chapter.
2. V. Perrone Compagni, ed. Cornelius Agrippa, *De Occulta Philosophia Libri Tres,* Studies in the History of Christian Thought, vol. 48, "Introduction" (Leiden: Brill, 2006), 3.
3. Three editions of the incomplete version appeared in 1531. The first was printed in Antwerp by Johannes Graphaeus and the second in Paris by Christianus Wechelus. Compagni, *De Occulta Philosophia Libri Tres*, 8.
4. The first edition was printed by Johannes Soter in Cologne, but the printing was suspended by the city in response to the Dominican inquisitor Conrad Kollin of Ulm, who decried the book as as "doctrina haereticus et lectione nefarius." Agrippa attempted to save his book with a letter to the senate justifying his work as part of a recognized and accepted tradition within natural philosophy, but the letter was highly critical of the senate and nearly scuttled the whole project. His patron, Archbishop Hermann von Wied, eventually stepped in to save the edition that appeared in July 1533 without the printer's name or place of publication and included in the table of contents sections of *De Vanitate* that attacked magic. Compagni, *De Occulta Philosophia Libri Tres*, 10. Subsequent partial editions were printed in 1567 in Paris by I. DuPuys and the team of Godefroy and Beringen, both of whom only printed book three. In 1650/1651, Gregory Moule printed John French's English translation in London.
5. Two editions of the fourth book were produced simultaneously in London. *Henry Cornelius Agrippa's Fourth Book of Occult Philosophy. Of Geomancy. Magical Elements of Peter de Albano: astronomical geomancy: the nature of spirits: and Arbatel of magic* (London: J.C. for John Harrison and J.C. for Matthew Smelt, 1655). Another edition with the same title was printed in London in 1783, and a German translation was printed in Stuttgart by J. Scheible in 1855. It is arguably the eighteenth-century edition that caught Mary Shelley's attention and caused her to employ Agrippa's magical work as the text that corrupted Victor Frankenstein.
6. Charles Nauert, *Agrippa and the Crisis of Renaissance Thought* (Urbana: University of Illinois Press, 1965).
7. Frances Yates, *Giordano Bruno and the Hermetic Tradition* (Chicago and London: Routledge and Kegan Paul, 1964).
8. Paola Zambelli, "Cornelio Agrippa nelle fonti e negli studi recent," *Rinascimento* 8 (1968): 173–176.

9. Compagni, *De Occulta Philosophia Libri Tres*, 16.
10. Christopher Lehrich, *The Language of Demons and Angels* (Leiden: Brill, 2003).
11. Compagni, *De Occulta Philosophia Libri Tres*, 2.
12. Keefer explores the importance of hermetic rebirth in Agrippa's philosophy, arguing that it was the means Agrippa used to help magicians understand the divine and the rewards to be had from this understanding. M.H. Keefer, "Agrippa's Dilemma: Hermetic Rebirth and the Ambivalences of *De Vanitate* and *De Occulta Philosophia Libri Tres*," *Renaissance Quarterly* 41 (1988): 614–653.
13. Moule, 3:4, 352. "*Religio ipsa est divinorum assidua comteplatio piisque operibus cum Deo divinisque numinibus religatio.*" Compagni, *De Occulta Philosophia Libri Tres*, 409. All quotations are taken from Heinrich Cornelius Agrippa von Nettesheim, *De Occulta Philosophia Libri Tres*, trans. John French (London: Gregory Moule, 1651). The footnotes reflect the 1533 edition printed in Cologne as edited by Compagni, *De Occulta Philosophia Libri Tres*.
14. Aristotle, *Metaphysics*, 6:1025b, ff, esp. 1026a: 18–19.
15. Moule, 3:8, 363. "*Hinc apud Plutarchum legitur gentiles deum decripsisse esse ipsum spiritum intellectualem et ingenum, non habentam formam, sed transformantem se in quodcunque voluerit et coaequantem se universis.*" Compagni, *De Occulta Philosophia Libri Tres*, 420–421.
16. Moule, 1:5, 10. "*Superni ignis propria sunt calor omnia foecundans et lux omnibus vitam tribunes; inferni ignis propria sunt ardor omnia consumens et obscuritas cuncta sterilitate complens. Fugat itaque coelistis et lucidus ignis tenebrosos daemones, fugat et hic noster ligneous eosdem, quatenus habet simularchrum et vehiculum lucis illius superioris; quin et illius qui ait: "Ego sum lux mundi," qui verus ignis est, pater luminum, a quo omne datum optimum venit, splendorem ignis sui emittens et communicans primo Soli et caeteris coelistibus corporibus et per haec, tanquam per media instrumenta, illum in hunc ignem nostrum influens.* Compangi, *De Occulta Philosophia Libri Tres*, 92–93.
17. Moule, 1:xi, 26. "*Accedit ad haec quod totidem sunt in anima mundi rationes rerum seminales quot ideae sunt in mente divina; quibus ipsa rationibus aedificavit sibi in coelis ultra stellas etiam figures impressitque his omnibus proprietates. Ab his itaque stellis, figures ac proprietatibus omnes specierum inferiorum virtutes et proprietates dependent, ita ut quaelibet species habeat figurum coelestem sibi convenientem, ex qua etiam provenit sibi mirabilis potestas in operando, qualem per rationes animae mundi semalies propriam ab idea sua sucsipit dotem.*" Compagni, *De Occulta Philosophia Libri Tres*, 107.
18. Moule, 3:8, 362. "*Divina enim mens, cum se unico et nunquam interrupt actu... prolem et Filium in se progenerat, quae est plena intelligentsia, plena sui imago et plenum mundi exemplar... Mercurius quoquoe Trismegistus in Asclepio Dei filium diversis in locis affirmat; inquit enim: "Deus mens atque pater mentem sibi aliam opificem peperit... Et in Pimandro, ubi vaticinari videtur de future lege gratiae ac de regenerationis mysterio, ait:*

"Regenerationisautor est filius Deo homo, unius voluntate Dei"; vocat etiam Deum itrusque sexus foecunditate plenissimum." Compagni, *De Occulta Philosophia Libri Tres*, 410.

19. Moule, 2:2, 171. "*Sunt itaquw numeri magnarum sunlimiumque virtutem potentes; neque enim mium est—cum in rebus naturalibus sinc tot ac tantae virtutes occultae,licet manifestarum operationum—esse in ipsis numeris multo quidem maiores, occultiores, mirabiliores atque efficaciores, separatis substantiis immixti, denique maximam et simplicissimam habentes cum ideis in mente divina commixtionem, a quibus propria et efficacissimas vires sortiuntur. Quamobrem etiam ad daemonica et divina munera consequenda plurimum valent plusque possunt, quemadmodum in rebus naturalibus qualitates elementales in transmutando ad aliquod elementale multum valent atque possunt."* Compagni, *De Occulta Philosophia Libri Tres*, 252.
20. Moule, 3:38, 363. *"Ex hac plena summae foecunditatis intelligentia amor producitur, ligans, ligans intelligentiam cum mente—et eo amplius, quanto immense proportione sibi intomor est quam aliae proles cum eorum parentibus—quae est tertia persona, scilicet Spiritus Sanctus."* Compagni, *De Occulta Philosophia Libri Tres*, 421.
21. Moule, 3:8, 364. *"Est igitur Deus, ut inquit Paul, "a quo omnia, in quo omnia et per quem omnia": a Patre enim tanquam a primo fonte emanant onmia; in Filio vera tanquam in piscine suis ideis collocantur omnia; per Spiritum Sanctum vero explicantur omnia et propriis gradibus singula distribuuntur."* Compagni, *De Occulta Philosophia Libri Tres*, 422.
22. Moule, 1:1, 1. *"Cum triplex sit mundus, elemtalis, coelestis, et intellectualis, et quisque inferior a superiori regatur ac suarum virium suscipiat influxum ita ut ipse Archetypus et summus Opifex per angelos, coelos, stellas, elementa, animalia, plantas, metalla, lapides, Suae omnipotentiae virtutes exinde in nos transfundat."* Compagni, *De Occulta Philosophia Libri Tres*, 85.
23. Moule, 1:1, 2. "*non irrationablie putant magi no per esodem gradus, per singulos mundos, ad eundem ipsum archetypum mundum, omnium opificem et primam causam, a qua sunt omnia et procedunt omnia, posse conscendere: et non solum his viribus quae in rebus nobilioribus praeexistunt frui posse, sed alias praetera nova desuper posse attrahere*. Compagni, *De Occulta Philosophia Libri Tres*, 85.
24. This kind of thought is visible in other Neoplatonic natural philosophers' work, including Iamblichus's. Agrippa just extended it in new ways to contend that magi could access the powers available at the creation to recreate the world.
25. Moule, 1:2, 3. "*phsyica docet naturam eorum quae sunt mundo illorumque causas, effectus, tempora, loca, modos, eventus, integritates et partes investigat atque rimatur.*" Compagni, *De Occulta Philosophia Libri Tres*, 86.
26. Moule, 1:2, 3–4. *"Mathematica vero docet nos planam et in tres porrectam dimensiones naturam cognoscere motusque ac coelestium progressus suspicere."* Compagni, *De Occulta Philosophia Libri Tres*, 87.
27. Moule, 1:2, 4. "*Theologia autem quid Deus ipsa docet, quid mens, <quid intelligentia, quid angelus, quid denique daemon, quid anima, qui religio,*

quae sacra instituta, ritus, phana, observationes sacraque mysteria; instruit quoque de fide, de miraculis, de virtute verborum et figureum, de arcanis operationibus et mysteriis signaculorum et, quod ait Apuleis, docet nos "rite scire atque callere leges ceremoniarum, fas sacrorum, atque ius religionum." Compagni, *De Occulta Philosophia Libri Tres*, 88.

28. Moule, book two, dedication, 166. *"in quo coelistis magfiae mysteria intimamus."* Compagni, *De Occulta Philosophia Libri Tres*, 247.
29. Moule, 2:1, 170. *"Inter omnia autem matematica numerous, ut sunt formaliores, ita etiam esse actualiores; quibus virtutem et efficentiam tam ad bonum quam ad malum non solum ethnicorum philosophi, sed etiam Hebraeorum ac Christianorum attribuunt theologi."* Compagni, *De Occulta Philosophia Libri Tres*, 251.
30. Moule, 2:2, 171, *"denique maximam et simplicissimam habentes cum ideis in mente divina commextionem, a quibas proprias et efficacissimas vires sortiuntur."* Compagni, *De Occulta Philosophia Libri Tres*, 252.
31. Moule, 2:7. 183. *"Quarternarium numerum Pythagorici tetractyn appellant ipsumque omnibus numerorum virtutibus praferunt,, <quai ipse fundamentum et radix omnium aliorum numerorum; unde et omnia fundameta tam in artificialibus, quam in naturalibus et divinis quadrata sunt, ut infra ostendemus; significatque soliditatemquae per figuram quadrariam etiam demonstrator. Est enim quaternaries quadrates planus primus qui constat duabus proportionibus, quarum prior est unius ad duo, posteror duorum ad quatuour; nasciturque gemina processione et proportione, videlicet unius ad unum et dorum ab duo, ab unitate incipiens, in quaternitate desines: quae proportiones in eo different quod iuxta arithmeticam sunt sibi in aequales, iuxta gemetrica vero aequales. Itaque quadratus Dompatri adscribitur, quin et totius trinitatis mysterium complectitur."* Compagni, *De Occulta Philosophia Libri Tres*, 263.
32. Ian Maclean details the multitude of techniques adopted by occult philosophers to both share and protect their secret knowledge. Based on a single excerpt from Agrippa's introduction to *De Occulta Philosophia Libri Tres* and another from Skalich de Lika's *Occulta occultorum occulta*, Maclean condemns the genre to eternal opacity, "This double gesture guarantees the survival of the genre by its forever deferred promise of explanation and clarity and its inbuilt need to itself be subjected to the exegesis that it had inflicted on the stemma of texts that precedes it." Ian Maclean, "The Interpretation of Natural Signs: Cardano's *De subtilitate* versus Scaliger's *Exercitationes*," in Brian Vickers, ed. *Occult and Scientific Mentalities in the Renaissance*, (Cambridge: Cambridge University Press) 1984, 231–252.
33. Don C. Skemer, *Binding Words: Textual Amulets in the Middle Ages* (State College, PA: Pennsylvania State University Press), 2006.
34. Moule, 3:10, 367. *"Opportet ergo scire sensibiles propretates per viam scretioris analogiae perfecte intellectualizare quicunque hymnos Oprhei et veterum magorum sane intelligere debeat: ita nanque comperiet haec ab cabalisticis arcanis orthodixisque tradiontibus minimum differe. Nam quos Curetes et intemeratos deos vocat Orpheus, Dionysius Potestates appellat, cabalistae illos numerationi Pahad, hoc est "timori divino," appropriant; ita quod Ensoph in*

cabala, Orpheus Noctem vocat; idem est Typhon apud Orpheum, qui Zamael in cabala." Compagni, *De Occulta Philosophia Libri Tres*, 424.

35. Moule, 3:11, 372. "*Atque haec sunt arcana abscondita, de quibus difficilimum est adferre iudicium et completam trader scientiam: nec in ulla alia lingua quam in hebraica intelligi et doceri possunt. Verum, cum divina nomina (ut ait Plato in Cratilo) habita sunt a barbaris qui habuerunt a Deo, sine quo nulletanus quis capere potest verso sermons veraque nomina quibus Deus nuncupatur, ideo de ipsis alia dicere non possumus, nisi quae a Deo sunt nobis sua benignitate revelata.> Sunt enim divinae omnipotentiae sacramenta atque vehicular non ab homnibus, non etiam ab angelis, sed ab ipso summo Deo certo modo pro suorum characterum immobile numero et figura sempiterna stabilitate institute divinitatisque harmonium spirantia divinaque assistentia consecrata: quare illa timent superi, tremunt inferi, colunt angelis, pavent cacodaemones et <omnis reveratur creatura>, omnis venerator religio. Quorum religiosa observation devotaque cum timore ac tremor invocatio virtutem nobis magnam praestant deificamque unionem atque estiam supra naturam mirabilium operum effectuumque potential.*" Compagni, *De Occulta Philosophia Libri Tres*, 430.
36. Moule, 3:7, 358. "*necesse igitur est magum quemque ipsum Deum, omnium productorem et primam causam caeterosque deos sive numina, quas secundas causas dicimus, cognoscere et quo cultu, qua veneration quibus sacris uniuscuiusque conditioni conformibus colendi sunt singuli non ignorare.*" Compagni, *De Occulta Philosophia Libri Tres*, 413.
37. Moule, 3:12, 380. "*Ab angelis itaque suscipiunt coeli quod influent, illi autem a magno nomine Dei et Jesu, cuius virtus prima est in Deo, post diffusa in duodenos et septenos illos angelos, per quos extenditur in dodecim signa et semptem planetas et consequenter in omnbes caeteros ministros et instrumenta Dei.*" Compagni, *De Occulta Philosophia Libri Tres*, 436.
38. Moule, 3:16, 390–394. Compagni, *De Occulta Philosophia Libri Tres*, 445–449.
39. Ibid.
40. Moule, 3:16, 393–394. "*angelos scelerum <et impietatis> ultores <juxta justitiae divine santionem>*". Compagni, *De Occulta Philosophia Libri Tres*, 449.
41. Moule, 3:20, 405–406; Compagni, *De Occulta Philosophia Libri Tres*, 459–460.

CHAPTER 3

The Astrological Cosmos of Johannes Kepler

Sheila J. Rabin

Johannes Kepler did not develop a traditional cosmogony, but he had an overriding idea about the creation of the universe that was fundamental to his view of the universe: "Geometry," he wrote in his 1619 treatise *The Harmony of the World*, "which before the origin of things was coeternal with the divine mind and is God himself..., supplied God with patterns for the creation of the world..."[1] This concept of the centrality of geometry to the divine plan of creation shaped Kepler's astronomical thought and undergirded his most notable contributions to astronomy, his laws of planetary motion, and his celestial physics. It shaped his very distinctive astrological thought as well.

Combining the ideas of Copernicus and Tycho Brahe, Kepler concluded that the universe was physical. That created a problem for his acceptance of astrology. Traditionally, belief in astrology had been grounded in the Aristotelian distinction between a physical sublunar world that was changeable and a celestial world that was eternal and immutable and moved the lower world. The celestial world could thus guide what happened on earth. The notion of a physical universe meant that the celestial world lost its superiority and consequent ability to guide us. Given his realization that the universe was physical, Kepler could have given up astrology altogether, but he was convinced that he had had successful experiences with it that reinforced its validity. In his 1606 treatise *On the New Star,* he stated that "by right, I should have been able to be seen as conceding everything to [Giovanni Pico della Mirandola's] judgment about the worthlessness of astrology";[2] he added, "I will not deny... that there is great vanity of experience vaunted by astrologers ..., but I will not on that account concede that experience has been nothing."[3] Kepler maintained that it was through astrology that he had accurately predicted a peasants' uprising and a Turkish invasion in 1595. He more often than not forecast the weather accurately, again using astrology. He believed he knew the treacherous Count Albrecht Wallenstein through the count's birth chart. More importantly, as J.V. Field has noted, "No general refutation of astrology was a reasonable proposition while the influence of the Sun upon the weather (in determining the seasons) and that of the Moon upon the sea (in causing the tides) were regarded as examples of astrological 'force' in action."[4] In Kepler's mind, astrology had proved and continued to prove itself. Therefore, he set about to reform it, and he did so in keeping with his belief that the divine creation followed geometrical principles.

In his reform, Kepler focused on aspects, the angles formed when the lines drawn from two planets, including the sun and the moon, converge on earth. Kepler called the aspects "a pearl of nobility from astrology,"[5] and while he did not invent the use of aspects in astrology, he was the first to suggest that they should be used to the exclusion of most other astrological mechanisms. Ptolemy had established four aspects: the conjunction, 0°; the square, 90°; the trine, 120°; and opposition, 180°. Kepler added three new aspects: the quintile, 72°; the sesquiquadrate, 135°; and the biquintile, 144°. Kepler used these aspects to forecast the weather, and he kept records to test the accuracy of his predictions. Kepler believed that looking at aspects could give insight into the human being as well. He suggested in his 1610 treatise in defense of astrology, *Tertius interveniens* (which, following Edward Rosen, I cite as *Third Man in the Middle*[6]), "The human being in the first igniting of his life, when he first lives for himself and cannot remain any more in his mother's body, receives a character and image of all the constellations of the heavens, or of the shape of the rays streaming toward earth and retains it until he is in his grave." Kepler maintained that this image of the heavens left an impression on the physical body of the person as well as on the character and personality, and it affected the person's relationship with other people as well. He concluded that by these means, "a very big difference between people will be produced, that one will become good, lively, joyful, trusting, another sleepy, indolent, careless, obscurantist, forgetful, timid, and what are such general qualities that are compared to the beautiful and exact or extensive, unsightly configurations and to the colors and movements of the planets."[7] For Kepler, the geometrical configuration of the heavens was imprinted on the human soul at birth. As the relationship of the planets to each other through their aspects caused certain personality traits to develop, the birth chart would give a description of a person's character. In this way, geometry was the archetype of the human being just as Kepler believed it was for the natural world. But planetary aspects are observable and measurable. Using Henry Cornelius Agrippa's definition of "occult qualities" that "their causes lie hid, and man's intellect cannot in any way reach and find them out" Kepler's astrology eschewed "occult" actions of the heavenly bodies[8] on earth and concentrated on overt, measurable mechanisms.

Kepler considered most of astrology superstitious and unacceptable. He outright rejected such common astrological doctrines as the distribution of the signs of the zodiac among the seven planets.[9] He rejected the possibility predicting the fortune and misfortune of the whole world, a country, a city, and so on, for he insisted that no one could establish a nativity for such entities; he also rejected a division of countries among the zodiacal signs.[10] Though he worked with houses and constellations of the zodiac, he saw their usefulness differently from other astrologers. He wrote in *On the New Star* that the images ascribed to the zodiac were not formed by nature; what they provided instead was a geometrical division of space.[11] The images of the zodiac were a source of sympathetic magic in astrology, as the qualities of the signs were believed to effect changes in the qualities of their subjects. Here, too, Kepler, dissociated his ideas about astrology from the occult. This could well have been the case when he rejected assigning parts of the human body to signs in medical practice.[12] Moreover, as he agreed with his adversary in *Third Man in the Middle*, the fixed stars do not move, and "because most of them stay in one place, their effect also remains basically one and the same; there is no noticeable difference."[13] The constellations in the zodiac could not form different aspects the way planets could; therefore, they could not indicate change.

However, Kepler accepted the astrological doctrine of progressions, the belief that the location of a planet at birth portended the character of the year in a person's life corresponding to the degree the planet has advanced from the subject's nativity. "The doctrine of progressions will earn fine consideration from me," he wrote in *Third Man in the Middle*. "If I would allow with Copernicus that the earth revolves, then the proportion naturally embedded between a day and the year turns out to be one to 365, whether we will be carried around in the universe with a field, a house, or a ship for our dwelling. And it is, therefore, more believable that in progressions and the nativities of human beings who are the inhabitants of this ship, this proportion should also rule."[14] Kepler found in the doctrine of progressions a natural geometric proportion that did not exist in other astrological doctrines, like the distribution of the twelve signs among the seven planets. Once again, a basis in geometry helped Kepler decide which theories to accept and which to reject. And he mentioned that he also liked the doctrine of progressions because he saw it as amenable to the Copernican system of a moving earth.

The idea that geometric proportionality was the foundation of what to accept in astrology led Kepler to further develop his theory of universal harmonies that found final expression in his work *The Harmony of the World*. According to Kepler, proportions from regular polygons, or knowable figures, as he called them, were harmonic. He outlined his thoughts on the issue in *Third Man in the Middle*: "[N]ature does not take pleasure from any proportion that would be taken from such rejected figures," he suggested,

> whether it be in voices or in rays of stars. And, on the other hand, all proportions of voices and chords which are taken from the knowable figures give in music its harmonies and in planetary rays all proportions that appear when two light rays strike together (as far as they are noted in the daily experience and recording of the weather itself in nature's drive toward violent weather), such [harmony] is also found under the knowable figures and not one under the unknowable. And so a wonderful secret follows from this, that nature is God's image and geometry is the archetype of universal beauty. So much was put into the work through creation; so much could be known in geometry through finitude and equations. And what falls outside the limits, comparison, and knowledge would also remain unformed and uncreated in the world. That which is given no special beauty or shape but of corporeality, fortune, and accident, which in themselves are boundless, would be abandoned as, for example, individual fruits and flowers are, indeed, found which have seven, nine, or eleven branches or leaves when the species commonly varies in the individuals. But no species is found which does not regularly contain this number, as five, six, four, three, ten, twelve, etc.[15]

For Kepler, here again geometry was the key for understanding all that is beautiful and true as the divine created the universe according to knowable geometric patterns. Violation of the natural geometric proportions resulted in discord, anomaly, and ugliness. Kepler believed this to be true in music; he believed it true in nature. In astrology, an aspect would produce harmony if it belonged to one of the regular polygons, discord if not. This was Kepler's view of the natural world: ordered, comprehensible, reducible to a single principle, as the divine creator's world ought to be. Fortunately for Kepler's formulation, the seven-toed ichthyostega was not discovered by the West until the twentieth century.[16]

Kepler believed in astrology, but he was not an astrological determinist; he did not believe that an astrologer could read the heavens to predict all future contingencies through astrology. "I do not mean to defend the prediction of future events that are contingent on the particular, for they depend on human free will," he asserted.[17] And he used a widespread prejudice to illustrate the futility of this belief: "The conjunction of Saturn and the moon should be the cause that someone is going to be cheated by a Jew. But if this conjunction takes place on the Sabbath, then no one in Prague will be cheated by any Jew, and, on the other hand, several hundred Christians will daily be cheated by Jews and vice versa, and yet the moon runs below Saturn only once a month."[18] Thus, human free will could interfere with the effects of the stars, making prediction through astrology not nearly as accurate as prediction through astronomy.

Though Kepler did not consider astrological prediction to be as reliable as astronomical prediction, like most of his contemporaries he did see astrology as an integral part of the study of the heavenly bodies. This is shown very clearly in *Third Man in the Middle.* In this work Kepler placed himself between two noted medical practitioners: Helisaeus Röslin, physician-in-ordinary to the Count-Palatine of Velderiz and the Count of Hanau-Lichtenberg in Buchsweiler and an astrological determinist on one side, and Philip Feselius, physician-in-ordinary to the Margrave of Baden, who believed astrology should be banned, on the other. Kepler had published an earlier tract against Röslin entitled *Antwort auf Röslini Diskurs* (Response to Röslin's *Discourse*) in 1609. *Third Man in the Middle*, published in 1610, was more specifically directed against Feselius. In it, Kepler consistently chided Feselius for his lack of knowledge about astronomy and mathematics, and the work is almost as much a defense of Keplerian astronomy just after the publication of the *New Astronomy* as it is a defense of astrology. In fact, Kepler further refined the celestial physics of the *New Astronomy* in *Third Man in the Middle*, for in the latter work he defined the mechanism that moved the planets as physical. Whereas for most examples of *species immateriata* in section 26 he used the Latin phrase, once he used German, and he called it *materialischer Außfluß*, a material emanation.[19] That Kepler should first make such an important physical-astronomical point in a defense of astrology suggests how closely astronomy and astrology were allied in Kepler's mind, that he saw astrology as a part of the study of the cosmos along with astronomy.

Kepler's rejection of doctrines in traditional astrology that did not conform to his geometrical conception of the universe also reflects an attempt to make astrology conform to his physical astronomy. The celestial configurations affected the earth and its inhabitants not only through geometry but also through their ability to alter the levels of moisture and heat.[20] This was particularly important in Kepler's weather predictions. But as much as Kepler tried to make astrology conform to his physical astronomy, the immaterial still operated in his astrology in a way that it did not operate in his astronomy. He believed that the heavenly configurations affect the weather because they act upon the earth's soul; the heavenly configurations affect human actions and characters because they act upon human souls.[21] On the one hand, Kepler's belief in the efficacy of souls in astrology allowed him to maintain the physical and geometrical unity of the cosmos; on the other, it meant, ironically, that despite his creation of a celestial physics, perhaps his most important contribution to the history of astronomy, Kepler's universe was still animate.

Kepler did not gain any disciples among students of the celestial sciences with his focus on the divine creator as a divine geometer in his cosmogony.

But his belief that geometry supplied the divine with the patterns of creations was one of his most fruitful ideas. From it came the impetus to his most important discoveries.[22] As an astronomer trained in the sixteenth century, Kepler was also taught to accept astrology as the "practical" side of the study of the heavens, and his practice of astrology reinforced his acceptance of its place in that study. But he could not integrate his acceptance of astrology with his Copernican astronomy, and so he set out to reform astrology with the same focus on geometry as the inspiration for the divine creation that was the foundation of his astronomy and cosmology. He rejected many traditional astrological ideas, including the belief that the images of the zodiac have a basis in nature and can effect changes. In keeping with his idea of the geometrical divine plan, Kepler emphasized facets of astrology that not only occur in nature but are also measurable. By emphasizing those points in astrology that were observable and measurable, Kepler was attempting to change the foundation of astrological thought by making it an open rather than an occult discipline. But just as was the case with his astronomy, that "Kepler's work was far too original to be absorbed readily" by astronomers,[23] so, too, was his astrology "far too original" to be understood and absorbed by astrologers.

Notes

1. Johannes Kepler, *The Harmony of the World*, trans. E.J. Aiton, A.M. Duncan, and J.V. Field (Philadelphia: American Philosophical Society, 1997), 304; *Gesammelte Werke* (hereinafter *GW*), ed. Max Caspar et al., vol. 6, *Harmonice mundi*, 223: "Geometria ante rerum ortum Menti divinae coaeterna, Deus ipse (quid enim in Deo, quod non sit Ipse Deus) exempla Deo creandi mundi suppeditavit, et cum imagine Dei transivit in hominem."
2. Johannes Kepler, *De stella nova*, in *GW* 1, 184: "... jure videri possim totus in ejus [Pici Mirandolae] sententiam de Astrologiae vanitate concedere." Translations are by me unless otherwise noted.
3. Ibid., 184: "Non nego magnam esse vanitatem experientiae ab Astrolgis jactatae...; at non ideo concedo, nullam fuisse experientiam."
4. J.V. Field, "A Lutheran Astrologer: Johannes Kepler," *Archive for History of Exact Sciences* 31 (1984): 220.
5. Kepler, *Tertius interveniens*, in *GW* 4, 209: "...ein Edels Perl auß der *Astrologia* ist..."
6. Edward Rosen, "Kepler's Attitude toward Astrology and Mysticism," in Brian Vickers, ed. *Occult and Scientific Mentalities in the Renaissance* (Cambridge: Cambridge University Press, 1984), 253–272, especially 257.
7. Ibid., 209–210: "...der Mensch in der ersten Entzündung seines Lebens/ wann er nun für sich selbst lebt/unnd nicht mehr in Mutterleib bleiben kan/ einen *Characterem* und Abbildung empfahe *totius constellationis coelestis, seu formae confluxus radiorum in terra*, und denselben biß in sein Grube hienein

behalte...ein sehr grosser Unterscheidt unter den Leuten gemacht wirdt/ daß einer wacker / munder / frölich / trauwsam: Der ander schläfferig / träg / nachlässig / liechtscheuh / vergessentlich / zag wirdt / und was dergleichen für general Eigenschafften seind / die sich den schönen und genauwen oder weitschichtigen unformlichen *figurationibus*, auch gegen den Farben und Bewegungen der Planeten vergleichen."

8. Henry Cornelius Agrippa, *Three Books of Occult Philosophy or Magic*, in *Complete Occult Philosophy Containing All Four Books*, anon. trans. (Whitefish, MT: Kessinger Reprint, 2005), 60.
9. Kepler, *Tertius interveniens*, in *GW* 4:185.
10. Ibid., 241–242.
11. Kepler, *De stella nova*, in *GW* 1, 168ff.
12. Kepler, *Tertius interveniens*, 226.
13. Ibid., 177. "Dann weil der meinste Haff an einem Ort stillsteht / so bleibt es auch mit ihrer Wirckung an und für sie selbst immer nur in einem: und gibt keinen mercklichen Unterscheidt."
14. Ibid., 185: "...wil bei mir die *doctrina directionum* ein feines Ansehen gewinnen / wann ich mit *Copernico* die Erde umbgehen lasse / dann als-dann findet sich die *proportio diei ad annum, hoc est, vnius ad* 365. unserm *domicilio* Hütten / Wohnung oder Schiff / darinnen wir in der Welt her-umb geführt werden / natürlich eingepflantzet: Und ist derohalben desto gläublicher / daß *in directionibus* und Natiuiteten der Menschen/welche dieses Schiffs Innwohner seind / diese Proportz auch regiren solle: als dann die *Astrologi* lehren."
15. Ibid., 204: "...die Natur sich ab keiner Proportz erfreuwet / die auß solchen verworffenen Figuren genommen were / es sei jetzo in *vocibus*, oder in *radiis stellarum*. Und hingegen / daß alle *Proportiones vocum seu chordarum*, die auß den *figuris scibilibus* genommen seind/*in Musica* ihre *concordantias* geben/und daß *in radiis planetarum* alle *proportiones*, die da erscheinen bei zusammenfallung zweier Liechtstraalen / (so fern sie tägli-cher Erfahrung unnd Auffschreibung deß Gewitters sich in Antreibung der Natur zu hefftiger Witterung mercken lassen) solche auch unter den *figuris scibilibus*, unnd kein einige sich unter den *non scibilibus* finden läs-set: Und also hierauß ein wunderbarliches *arcanum* folget/ daß die Natur Gottes Ebenbildt/und die *Geometria archetypus pulchritudinis mundi* seie/ darinnen durch die Erschaffung so viel ins Wercke gestellet worden / so viel *in Geometria per finitatem et aequation / es* müglich gewest zu wis-sen / unnd was ausserhalb den Schrancken der Endtlichkeit Vergleichung und Wissenschafft gefallen / dasselbige auch in der Welt ungeschaffen und ungemacht geblieben seie: Das ist / keine besondere *pulchritudinem* oder *species* gegeben/sondern der Materialitet / *fortunae et casui*, die an ihnen selber unendtlich seind/uberlassen worden / als zum Exempel finden sich wol einzehle Früchte unnd Blumen / die sieben / neun /oder eilff Fäche oder Blätter haben / wann die *species in indiuiduis* gemeiniglich variert / aber kein *species* findet sich nicht / die diese Zahl beständig halte / wie fünff / sechs / vier / drei / zehen / zwölff/&c."

16. See Stephen Jay Gould, "Eight Little Piggies," in *Eight Little Piggies: Reflections in Natural History* (New York: W.W. Norton, 1993), 63–78.
17. Kepler, *Tertius interveniens,* in *GW* 4:198: "... ich nemlich nicht gensinnet / die vorsagungen *futurorum contingentium in indiuiduo,* so ferrn sie von deß Menschen freiem Willen dependirn / zu vertheidigen."
18. Ibid., 163: "... die *conjunctio Saturni et Lunae* Ursach gewest sein solle / daß einer von einem Juden betrogen worden ist. Dann wann diese *conjunctio* geschicht am Sabbath / so wirdt zu Prag niemandt von keinem Juden betrogen / unnd hingegen werden täglich etlich hundert Christen von Juden *et contra* betrogen/so doch der Mondt im Monat nur einmal zum *Saturno* läufft."
19. Kepler, *Tertius interveniens,* in *GW* 4, 170. See my article, "Was Kepler's *Species immateriata* Substantial?" *Journal of the History of Astronomy* 36 (2005): 49–56, regarding *materialische Außfluß* as a translation of *species immateriata.*
20. Ibid., 172–174. See also *De fundamentis astrologiae certioribus,* in *GW* 4, 16–17. English translation by Field in "A Lutheran Astrologer: Johannes Kepler," 239–242.
21. On souls in Kepler's astrological thought, see Patrick J. Boner, "Soul-Searching with Kepler: An Analysis of *Anima* in His Astrology," *Journal for the History of Astronomy* 36 (2005): 7–20. See also my article, "Kepler's Attitude toward Pico and the Anti-astrology Polemic," *Renaissance Quarterly* 50 (1997): 750–770, esp. 764.
22. On the value of the geometrical archetypes, see Rhonda Martens, *Kepler's Philosophy and the New Astronomy* (Princeton: Princeton University Press, 2000), esp. Chapter 4.
23. Bruce Stephenson, *The Music of the Heavens: Kepler's Harmonic Astronomy* (Princeton: Princeton University Press, 1994), 242.

CHAPTER 4

A Theater of the Unseen: Athanasius Kircher's *Museum in Rome*

Mark A. Waddell

It is indeed absolutely plain that all things seen by us are in truth other than what they seem.

—Athanasius Kircher, *The Great Art of Light and Shadow* (1646)

The celebrated museum of the Jesuit naturalist Athanasius Kircher (1602–1680) seems a particularly apt subject for a collection on world-building, for it was an entire world unto itself, an elaborate and bewildering cosmos that existed nowhere else in early modern Europe. In its heyday, the museum was an intellectual and spectacular marvel, visited by popes, cardinals, and princes as well as a host of intellectual luminaries from both sides of the confessional divide. A visit to Rome by the virtuosi of Europe inevitably included a sojourn in Father Kircher's wondrous collection, and Kircher himself once boasted in a letter that "no foreign visitor who has not seen the museum of the *Collegio Romano* can claim that he has truly been in Rome."[1]

For the historian, the collection that Kircher expanded and controlled in the Society's house of learning in early modern Rome, the *Collegio Romano*, offers an intriguing glimpse of his epistemology. It was a visual and tangible expression of the principles that governed his pursuit of knowledge; it was also inherently ephemeral—its fortunes were closely tied to Kircher himself and, as he declined, so too did his collection. It disintegrated swiftly following his death, in spite of the efforts to preserve it made by Filippo Buonanni in the early years of the eighteenth century.

Kircher was not, however, a simple collector of natural marvels. He was an active world-builder whose museum demonstrated his strong interest in how we come to know things about the universe. The Kircherian collection did more than display tantalizing hints of Nature's secrets, as some have suggested; it also displayed the methodological keys required to unlock those secrets. It trained its audiences in *how to look*, an exercise mediated in this case by the intervention of artifice, even as it pointed to the fallibility of our senses. Thus, the central core of the collection was concerned less with accumulation and more with querying the act of knowing.

At the same time, the collection also embodied important epistemological changes sweeping through the Society of Jesus in the seventeenth century. In its attention to the use of apparatus and artifice, the collection embraced the disciplines of mixed mathematics and, perhaps, anticipated the culture of machines that would come to prevail in the Royal Society under the auspices of Boyle, Hooke, and others. In its public character it emphasized the necessity for universal assent—still crucial to the Aristotelian process of knowing—while its crowded galleries of spectacular machines and philosophical puzzles called attention to the singular, subjective experience. It both challenged and celebrated the exercise of the senses, straddling what was becoming an increasingly fraught boundary between sensualism and skepticism. Most importantly, in its rhetorics of similitude, spectacle, and correspondence the collection encouraged imaginative and probable considerations of natural phenomena rather than offering certain and evident demonstrations of causes.

The Kircherian museum offers a fascinating window on the ways in which early modern Jesuits struggled to frame and understand the natural world. The cosmos presented by Kircher was a theater of the unseen, simultaneously a celebration and a revelation of the hidden parts of Nature. This almost paradoxical identity resulted in a space that functioned somewhere between spectacle and instruction, bemusement and enlightenment. Indeed, "bemusement" seems a good term to apply to this collection: Kircher wanted his audiences to be hesitant, overwhelmed, confused—in a word, *uncertain*. Only then could they be instructed in new and more powerful ways of understanding the world.

Situating Collection and Curator

Kircher was fond of embellishing the details of his early life (figure 4.1). He claimed to have narrowly escaped death as a young boy after falling into the local river and being swept under a mill wheel. Entering the Society of Jesus in 1618 at the age of sixteen, he was displaced repeatedly over the next several years by the constant warfare then plaguing Germany, and eventually found himself floating down yet another river on a piece of ice—with, he later claimed, a murderous mob of Protestants in pursuit. More Protestants beat and threatened to hang him, and his improbable survival, Kircher decided, was a clear case of divine providence. He reached Lyons

Figure 4.1 Athanasius Kircher, S.J., at the age of 62. From his /Mundus Subterraneus/ (Amsterdam, 1664). Courtesy of the Roy G. Neville Historical Chemical Library of the Chemical Heritage Foundation.

in 1631, and remained in southern Europe for the rest of his life. He was eventually appointed professor of mathematics at the *Collegio Romano* and, despite repeated requests that he be allowed to leave and travel abroad, Kircher remained in Rome for most of his later life, where he founded his famed museum at the *Collegio*, produced more than thirty published works, and established an extensive network of correspondents both within and beyond the Society.

The Kircherian collection was the direct descendent of another philosophical collection already housed in the *Collegio*, the "mathematical museum" of Christoph Clavius (1538–1612) and his student Christoph Grienberger (1561–1636), both of whom were instrumental in shaping the mathematics curriculum within the Society.[2] Kircher's collection took shape in the 1630s, housed originally in his own apartments in the *Collegio*, and in 1651 it was greatly expanded by the addition of another collection, that of the Roman patrician Alfonso Donnino, and was moved to a larger and more public space. At the time it was merely the latest in a long line of collections and theaters of nature located in the Eternal City, but its identity as a public showpiece for the Society of Jesus and, more particularly, for Kircher himself soon made it a unique and much-visited space in baroque Rome.

In some respects, the Kircherian museum was in fact typical of its time, with its juxtaposition of *naturalia*, antiquities, and artwork, but it was primarily in the element of bemusement that this museum differed from other, contemporary collections. While those other collections typically demonstrated the erudition or wealth of their owners as well as organized or systematized the world in a way that permitted investigation and understanding, the Kircherian museum, while advertising the resources and accomplishments of the Jesuit order, became primarily an expression of Kircher's unique worldview. He used his collection to portray the world as fundamentally mysterious, crisscrossed by invisible forces that could be revealed only by carefully orchestrated philosophical displays in which Nature was captured and made to work as Kircher himself directed. At the same time, however, he also used these displays to instruct his audiences in the fallibility of their senses, the ease with which their eyes could be tricked and their vision of Nature distorted.

Kircher himself has been described as his museum's most spectacular exhibit,[3] but in deliberately exploiting the bemusement of his audience he also fashioned for himself a powerful role as mediator and guide. In the production and consumption of spectacle there is always an exchange of power: a willingness on the part of the audience to reveal, and to revel in, their ignorance of the principles behind the spectacular displays, and a concomitant assertion of virtuosity and mastery on the part of the wonder-worker—mastery, that is, over the audience itself as well as over the physical principles that make his marvels possible. In wrapping his philosophical principles in spectacular displays and a playful extravagance, Kircher confirmed his own mastery even as he beguiled and instructed his visitors.

According to some historians, however, Kircher's virtuosity in his museum was centered more around the encyclopedic accumulation of objects and ideas than the active promulgation of his worldview. Paula Findlen has discussed the Kircherian collection at length, almost always emphasizing its encyclopedic character, as when she refers to the museum as an "illustrated encyclopedia"[4] whose "primary function . . . lay in the identification of *signs*."[5] She emphasizes "the accumulation of objects and publications" because it reinforces her characterization of Kircher as "the embodiment of a new form of expertise—not the 'on-site' knowledge of the traveler but the more synthetic knowledge of the collector whose wisdom surpassed the abilities of any one individual."[6]

Michael Gorman—who, with Findlen, has contributed the lion's share of our current understanding of the Kircherian collection—advances a parallel line of reasoning. In the context of his own discussion of Kircher's museum, he suggests that "one of the preconditions for the circulation of beliefs concerning wonders of nature and art is that one accepts one's source as being a passive conductor of information, a disinterested mediator of knowledge."[7] This view of Kircher as passive and disinterested dovetails with the courtly ideal of *sprezzatura* in which virtuosity and disinterestedness went hand in hand, and locates the Kircherian collection more firmly in that courtly culture.[8]

There is evidence, however, that Kircher saw his own role to be that of an active disseminator of ideas rather than a passive collector. His erstwhile disciple Giorgio de Sepibus confirmed as much in 1678 when he published a catalogue of the Kircherian collection: the *Romani Collegii Societatis Jesu Musaeum Celeberrimum*, or "The Most Celebrated Museum of the Roman College of the Society of Jesus." In its preface, De Sepibus concluded by saying, "The workshop of Art and Nature, the treasury of the mathematical disciplines, the epitome of philosophical practice—this is the Kircherian museum."[9] His use of the word "workshop" (*ergasterium*) implied activity rather than passivity, and the collection's identity as a "treasury" (*gazophylacium*) of the (mixed) mathematical disciplines was undoubtedly rooted in its culture of machines. More significant, however, was his use of the phrase "philosophical practice." Kircher did not envision his collection as a simple repository for objects but as, instead, a tangible demonstration of how best to practice natural philosophy. Its purpose was explicitly both epistemological and practical.

Another problem with the notion of a passive, encyclopedic Kircher is that there remains little role in such an interpretation for his audiences. This is a significant oversight, for Kircher must be acknowledged as a consummate manipulator of the baroque audience. His entire philosophical endeavor, both within his museum and in the texts that he published beyond its walls, depended absolutely upon the cooperation of spectators. The very purpose of the collection was to be witnessed by others; without the crucial act of seeing or observing, it was indeed merely an accumulation of objects. Findlen has suggested that the loadstone or magnet acted as the "key" to the Kircherian museum,[10] but I would suggest instead that the true key to the collection was its audience—in an important sense, its efficacy was made possible only by the attention of its visitors. Without them, the museum failed to work as intended.

This central role of the audience is communicated explicitly in the De Sepibus catalogue (figure 4.2). On its frontispiece, we see Kircher speaking with a pair of visitors, presumably greeting them as they enter; the letter with which he has been presented is probably an introduction. The tableau, and particularly the letter, are themselves expressions of the carefully regulated social conventions that dictated such encounters. It captures as well the important fact that it was Kircher who met with and guided his visitors. There were no unaccompanied strolls through the collection; as a result,

Figure 4.2 An idealized portrait of the Kircherian collection in the /Collegio Romano/, from the frontispiece to Giorgio de Sepibus? /Romanii Collegii Societatis Jesu Musaeum Celeberrimum/ (Amsterdam, 1678). Courtesy of Department of Special Collections and University Archives, Stanford University Libraries.

the museum provided an excellent opportunity for Kircher to present his idiosyncratic view of the world to a select but influential public.

The tableau presented by the De Sepibus catalogue reflects as well the reality of the Kircherian museum as a public space, and indicates that it was in its very "public-ness" that its purpose resided. In the Renaissance, collections were usually located in the private homes of their collectors, a trend that began to change in the sixteenth and seventeenth centuries.[11] Initially, of course, Kircher's own collection was confined to his apartments in the *Collegio Romano*, but following its expansion in 1651 it lost this sense of personal intimacy and became, instead, something larger, more public, and considerably more effective as a demonstration of the Society's resources and erudition. Indeed, the eventual fate of the collection testifies to the integral link between "public-ness" and purpose: toward the end of Kircher's life, his museum was reduced to occupying a dark and narrow corridor in the *Collegio*, a far cry from the vaulted, lavishly decorated gallery of its heyday. Thus, as the personality behind the collection waned along with its prestige, so too did its exposure; its utility was directly proportional to its public character.

Both audience and "public-ness" were, in fact, central to the collection's identity as "the epitome of philosophical practice," as De Sepibus phrased it. As a series of public demonstrations, the Kircherian collection embraced at least a nominally Aristotelian orientation, particularly in its attention to the line between singular and collective experiences. Traditionally, scholastic philosophers considered an explanation for natural phenomena "evident" only insofar as it met with collective assent, which created problems for scholastic thinkers interested in disciplines like astronomy and other examples of the mixed mathematical sciences.[12] Astronomical observations, for example, were always singular experiences conducted at specific places and times—the very antithesis of the "universal experience" demanded by conventional scholasticism. The decades leading up to the establishment of the Kircherian collection saw widespread debate within the Society of Jesus as some sought to reconcile the traditional demands of their shared philosophy with novel and increasingly important practices of observation. The result was a sort of middle ground, a series of strategies whereby Jesuits like Christoph Scheiner (1573–1650) and Giovanni Battista Riccioli (1598–1671) collected and collated singular experiences into something that approached universal assent.[13]

As we see below, many of the machines and puzzles that filled the Kircherian museum also encouraged the consideration of singular experiences. That these elaborate displays were exposed to public contemplation, however, transformed them into something closer to a collective or universal experience. The deliberate repetition of magnetic and optical devices within the collection accomplished much the same thing. Franciscus Aguilonius (1567–1617) and Louis Bertrand Castel (1688–1757), Jesuit philosophers who together bracketed Kircher's own time in the Society, both argued that "the repetition of many [singular] acts," each exhibiting the same behavior, could in itself create the conditions necessary for collective or universal assent.[14] This may explain, for example, why Kircher did not display just one or two magnetic machines; he displayed many of them, each individually demonstrating the basic behavior of the magnet but all of them collectively making that behavior "evident" in the philosophical sense. Thus, both the collection's public identity and the active participation of its audiences were integral to establishing its philosophical legitimacy, at least in an Aristotelian context.

The De Sepibus Catalogue

Atalgisa Lugli has suggested, intriguingly, that Kircher's extensive collection simply could not be contained entirely within the walls of his museum, and consequently spilled over into his published texts in a blurring of boundaries and spaces.[15] Indeed, I believe that at least part of Kircher's literary output should be understood as an example of the *museo cartaceo* or "paper museum" that was favored by early modern naturalists like Aldrovandi and Cassiano dal Pozzo. The world Kircher presented in his museum was not confined to the *Collegio Romano*; it existed as well in the pages of his books, with each acting as an extension of the other. This expresses forcefully both universality and epistemological coherence. We find in the Kircherian museum the same emphases on vision, speculation, and imagination that permeated his published texts; indeed, it is a relationship between presentation and knowing that seems better-suited to the museum than it does to the text. Perhaps the notion of the *museo cartaceo* is even more important than we might suspect at first glance. Kircher's general epistemology was closely allied to the rhetorics of spectacle, secrecy, and control that shaped his collection, and his museum may thus hold the key that unlocks much in Kircher's published works.

The catalogue compiled by Giorgio de Sepibus straddles, too, the boundaries between collection and text. More than anything else, it conveys an impression of abundance, a dazzling cornucopia of exotic animals, strange machines, and beautiful pieces of art. Undoubtedly, the museum was intended to awe its visitors with such tangible manifestations of Jesuit influence and wealth; the spiritual resources of the Society of Jesus were substantial, but their material resources were no less so, as signified by the tributes and *exotica* that flowed into Rome from the entirety of the world. The frontispiece of the catalogue certainly conveys this impression, depicting the museum itself as a beautiful and grandiose space, full of light and existing somewhere between elegance and outright clutter.

In the foreground, as we have seen, Kircher himself greets a pair of guests, an expression of humanist *politesse* that emphasizes the fact that a tour of the museum was inextricably linked with the presence of its administrator. The tiny human figures lend a sense of scale to the collection as well as to the space itself, making both appear truly spectacular. Findlen interprets this visual diminution of Kircher as a negative commentary on his role in his collection: "Bringing all of nature into the museum, mankind has ultimately been dwarfed in the process. Kircher does not control the objects in his collection; they control him."[16] I disagree, however. The museum depended absolutely upon Kircher's personal control; without his presence, the collection would—and ultimately did—disintegrate altogether. The museum may indeed have dwarfed its visitors with its staggering array of *naturalia*, technologies, and artistry, but the same cannot be said for its controller.

This point is made visually by the fact that the figure of Kircher, small though it is, stands in the very center of the foreground, his black robes forming a sharp contrast with the pale tiles behind him and thus drawing the eye of the viewer irresistibly to him. The grandeur of the depicted space, then, does not so much dwarf Kircher as bend itself around him; the purpose in depicting this grandeur is to glorify the collection and, by extension, the Society itself, rather than to diminish its curator.

Prominently displayed around Kircher and his guests are several obelisks. As De Sepibus tells us, there were six altogether, "four large and two small," and they marched along a promenade within the museum, their "Egyptian marks" clearly visible to visitors.[17] Along the walls, De Sepibus adds, were a total of thirty-four columns supporting various pieces of marble sculpture and statuary, and in the frontispiece these columns can be seen receding away down the central promenade, displaying carved busts in the classical style. To the far left of the image are shelves holding a variety of urns, presumably of ancient provenance, as well as a host of *naturalia*, including what appear to be the three rhinoceros horns mentioned at one point in the catalogue,[18] an elephant's tusk, and, hanging from the ceiling, the obligatory stuffed crocodile, one of which was displayed in practically every collection of natural curiosities in the early modern period (Kircher apparently owned two of them, with one sent all the way from Java).[19] A human skeleton towers in the shadows nearby, perhaps signifying both anatomical demonstration and the sober reflection of the *memento mori*. Paintings decorate the walls—De Sepibus tells us that they depicted popes, Holy Roman Emperors, and other princes of repute as well as important members of the Society of Jesus, including Christoph Clavius, "the incomparable mathematician"—and though obvious pieces of technology are largely absent from this image, a device at the far right, hanging next to a window, is probably some sort of thermometer (or, as De Sepibus called it, a "thermoscope").

This depiction portrays a fairly standard seventeenth-century collection, albeit on a grander scale than those of Cesi or Aldrovandi. Natural curiosities were placed next to antiquarian pieces, works of art, and examples of technology in a seeming hodgepodge that looks, to modern eyes, hopelessly confused. In fact, the Kircherian collection, with its exuberant intermingling of objects and categories, frequently blurred ontological and aesthetic boundaries as it brought together the natural, the artificial, and the supernatural in novel and surprising ways.

The frontispiece provides us with a visual approximation of the space occupied by the collection; the text that follows allows us to imagine that we have stepped into the museum ourselves. As we entered, we would have found to our right a library "stuffed [*refertum*] with the works of unique authors" that included a rare manuscript copy of Avicenna's treatise on plants in Latin, Hebrew, and Arabic. To the left was "an extensive revolving platform with many rare coins of twelve Roman emperors.... This is followed by a table absolutely crammed [*refertissima*] with lapidary arts, and with sculpture."[20] De Sepibus's choice of language—"stuffed with works," "absolutely crammed"—was surely significant, for in these first sentences he conveyed an impression not merely of abundance but, in fact, of there being almost *too many* things, of shelves and tables that were barely sufficient to contain and display their collections.

This impression is only strengthened by what follows—namely, some sixty pages of lists and descriptions, reeled off with a brisk efficiency. There are hydraulic and "hydrotechnic" machines, many of them clocks, as well as "a crystal globe full of water in which the resurrection of the Savior is represented in the middle of the waters." Also mentioned are an assortment of animals native to the Indies and Mexico, as well as a number of brightly colored birds from Brazil. There were examples of glass artistry and related objects, including four glass vials, "hermetically sealed," containing water from the River Jordan that had remained "uncorrupted" for some sixty years—not, perhaps, the most exciting exhibit in the museum—and a collection of "mathematical instruments" that included astrolabes, armillary spheres, and terrestrial globes.

Arriving next at the magnet, De Sepibus launched into a flurry of extravagant phrases, calling it "the prodigy of nature, thaumaturgical wonder, labyrinth of hidden and inscrutable virtues, the innermost part [*medulla*] of the Kircherian museum."[21] De Sepibus's rhetorical flourishes here undoubtedly expressed precisely the idea that Kircher sought to convey with his magnetic machines. The emphasis on "hidden and inscrutable virtues" and the reference to thaumaturgy both point to an atmosphere of mystery and obfuscation; what these machines offered to their audiences was not a sober and philosophical discussion of the magnet's properties but, rather, a lavish spectacle in which the magnet was cast as a prodigious marvel of Nature. Its virtues and properties were not necessarily exposed or revealed to the gaze of the curious visitor; instead, it was the magnet's mysterious power on display.

This rhetoric also supports Paula Findlen's claims that the magnet had a special significance for Kircher and, particularly, for his collection. Findlen describes the loadstone as demonstrating the invisible correspondences that webbed the universe and tied together disparate things, and this was a theme that Kircher explored in detail in his *Magnes; sive, De arte magnetica* of 1641 and that reappeared in later works as well. When De Sepibus called the loadstone the *medulla* of the museum, we are meant to understand that, though these magnetic machines take up only four pages of the catalogue, they played a far more significant role in the life of the museum. Indeed, given this language, we can probably assume that Kircher highlighted the presence of these devices when escorting his visitors through the collection; likewise, we can assume that the purpose Kircher envisioned for these machines was both different in kind and more significant than what he envisioned for his collections of exotic birds and Roman coins.

By all accounts, these machines were delightful expressions of baroque spectacle. Some were described specifically as instructing those who examined them—"This machine teaches you a new method of finding the meridian line" (*Docet te hæc machina novam praxin inveniendi lineam meridianam*)—while others were reiterations of well-known feats of ancient ingenuity, such as the wooden dove fashioned by Archytas that reputedly was capable of flight. There were models of the geocentric universe that rotated once every twenty-four hours thanks to the guidance of an "occult motion," and De Sepibus added that the collection boasted "not one, but three" reproductions of the famous planetarium of Archimedes mentioned by Cicero and other classical authors, displayed on the promenade in the middle of the museum (figure 4.3).

The common thread running through these devices was not simply that they were driven by magnetism, but that they were presented as embodying "secret forces" and "occult motions." Objects moved seemingly by themselves, as in the replication of a trick attributed to the mythological inventor Daedalus in which a statue was made to move "by its own will."[22] Daedalus supposedly had used quicksilver to animate a statue of Aphrodite and, while De Sepibus did not elaborate on the moving statue found in the collection, Kircher's *Magnes* of 1641 revealed that it was actually a statue of Daedalus himself that was made to glide across a stage by means of a magnet hidden beneath the floorboards, a sly and entirely Kircherian attempt at one-upmanship.[23]

The question that remains, however, is how Kircher sought to use these machines. It is clear that they were designed explicitly around insensibility and invisibility; their power and utility as spectacles was rooted in the secret and hidden nature of the loadstone and not in the philosophical revelation of magnetic properties and virtues. De Sepibus tells us repeatedly that these devices worked by means of an occult power or a hidden force—they were presented not as simple tricks explained by a hidden magnet, but as elaborate demonstrations of a secret, hidden thing.

This point is a crucial one, for these machines—collectively; the centerpiece or *medulla* of the Kircherian museum—celebrated rather than

Figure 4.3 The legendary planetarium of Archimedes, labeled here as "The Sphere of Archimedes, exhibiting the motion of the heavens by magnetic artifice." From Kircher?s / Magnes; Sive, De Arte Magnetica/, 2nd ed. (Cologne, 1643, p. 305). Courtesy of Department of Special Collections and University Archives, Stanford University Libraries.

exposed the insensible. At the same time, by deliberately focusing attention on what could not be seen, they diminished the importance of the senses in the study of nature. If the magnet and its occult power symbolized the correspondences that supported the very fabric of the Kircherian world, how could one study such correspondences when the properties of a simple loadstone defied observation?

By focusing the attention of his audiences on an invisible and intangible thing, Kircher deliberately engineered a situation in which the normal or conventional study of natural phenomena was frustrated and, ultimately, denied. If, as I suggested earlier, the Kircherian collection trained its audiences in *how to look*, then these machines offered an excellent opportunity to do precisely this: they challenged the conventional strategies of observation, making clear that a different means of observing these forces and phenomena was needed.

The method I believe Kircher wished to present with these machines revolved around the exercise of the imagination, whose importance had been increasing steadily both across European culture more generally and within Aristotelian psychology more specifically.[24] These magnetic devices demonstrated to their audiences that though the magnetic force could not be observed directly, it could be *imagined*. Where the senses failed, the imagination could be trained to thrive, spurred to do so by pleasing and lavish spectacles that forced the mind to "see" what the senses could not.

These devices were designed not to display or present particular philosophical problems, but to train their audiences in a kind of imaginative visualization that had already been well-established at the very core of Jesuit culture by the beginning of the seventeenth century.[25] The spectator was urged to move beyond the immediate impressions of his senses, to see more than magnets and to imagine instead concepts and things that transcended any single machine. Like the images commissioned as part of elaborate meditative texts circulating within the Society of Jesus, which provided the imagination and the mind with sensual spurs intended to elicit contemplation, the Kircherian machines presented in their elaborate workings not a visual demonstration of natural principles but, rather, the foundations of an imaginative exercise by which the spectator came to comprehend the structure and activity of the world.[26] They became the instruments through which one imagined and contemplated the unseen.

As I have explained elsewhere, Kircher was part of a culture in which the exercise of the imagination was not merely accepted, but was in fact encouraged and promoted by meditative and intellectual exercises of great complexity.[27] In creating and displaying elaborate mechanical spectacles that had as their subject an invisible and intangible thing, he was engaged in the same kind of activity as those instructors at Jesuit colleges who staged spectacular plays and tableaus designed to convey fundamental truths about the Catholic faith. In both cases, the goal was to encourage a meaningful contemplation of an invisible subject through the careful use of analogy, metaphor, and spectacle, thereby inculcating in the receptive mind an understanding of the subject.[28]

At the same time, if we do not accept that Kircher buried messages in these machines about how to study and conceptualize the world, we are left with a collection of devices that were effectively devoid of intellectual merit. In other words, without an avowedly epistemological purpose that commented on the act of knowing, they were philosophically useless. They cannot properly be considered natural philosophical instruments because they did not establish or demonstrate anything about the loadstone beyond its ability to move either pieces of iron or other loadstones—that was all these machines did, and surely Kircher did not need or want merely to rehearse this one basic property of the magnet some dozen times or more. Nor can they be considered philosophically instructive, for their audiences were patently *not* instructed in the causes behind the observed effects, which were presented instead as mysterious, hidden, and secret.

I am unwilling to believe that Kircher commissioned, constructed, and displayed a series of useless machines. Not only would this be wasteful, but it would also contradict blatantly the obvious messages and purposes articulated throughout his museum, which was presented as "the epitome of philosophical practice." Indeed, as I have already argued, the collection itself did more than merely capture and display natural phenomena; it instructed its audiences in how to study and interpret them as well, and it is in that spirit that I suggest an epistemological orientation for these magnetic machines. Moreover, this idea is bolstered, as we shall see shortly, by the messages conveyed by other machines in the Kircherian collection.

Before we turn to those other machines, however, I wish to draw a hasty comparison between Kircher's seeming purpose in the display of such devices and the rhetoric of René Descartes, who claimed in Part IV of the *Principles* that the use of artifice allowed one to witness and see the operations of Nature that were otherwise too small to be witnessed or that were altogether insensible.[29] As Brian Baigrie puts it, "The lesson that Descartes and his fellow mechanists extracted from the science of machinery was that insensible causes of natural phenomena can be 'seen'—in a matter of speaking—and therefore rendered intelligible by reconceptualizing these phenomena as systems of rigid parts that collaborate in the production of mechanical effects."[30]

Descartes went on to argue that both the senses and the imagination were useful only in representing corporeal things; they were more likely to deceive if brought to bear on incorporeal subjects. But for Kircher, and for Aristotelians in general, even the imagined impression of an intangible or incorporeal subject provided something from which the intellect could abstract a degree of comprehension.[31] Thus, it makes sense that the magnetic machines at the heart of the Kircherian collection could be used to translate the magnet's invisible power into a form of spectacular mechanical motion, creating a means of "seeing" this occult power at work in displays that themselves simulated the wider world. The attentive spectator thereby came to understand the behavior of these insensible forces as well as their role in the larger universe. Artifice was ideally suited to this revelation because it imitated nature—it translated natural processes into systems that were observable and controllable. Descartes argued this point explicitly, but I believe we must read the Kircherian collection as articulating this idea as well; in both cases, the artificial emulation of Nature was understood as an instrument of revelation, making insensible processes both visible and comprehensible.[32]

Labyrinths, Fallacies, and Subterfuge

Arguably the single greatest limitation to studying the Kircherian museum is that we are rarely, if ever, enlightened to the purpose of the machines and puzzles displayed therein. We know what they looked like, how they moved, sometimes even how they were constructed, but all too often that is all we know. My suggestion that the collection's magnetic machines were intended both to highlight the fallibility of the senses and to encourage the imaginative contemplation of insensible phenomena is, I think, a plausible explanation for the motivations that drove Kircher to construct and highlight these devices, particularly when placed in context with other machines in the museum that articulated related messages about seeing and knowing.

For example, as part of his discussion of "Optical, Catoptrical, [and] Dioptrical Experiments," De Sepibus described the *theatrum catoptricum*, which addressed the "much-discussed question among Physicists, *of Infinite action*."[33] The principle of the device was simple: a number of mirrors were arranged in such a way that any object placed in the *theatrum* was reflected to a seemingly infinite degree. The spectator presumably would then use this optical demonstration to contemplate the possibility of actions extended to infinity. This was by no means an idle question: the concept of infinity had been debated by scholastic thinkers in both theological and mathematical contexts from the medieval period, and clearly the question was, in the middle of the seventeenth century, still "much-discussed"[34] (figure 4.4).

Upon detailing how, by placing a few miniature trees or shrubs in the *theatrum* one could produce the appearance of an entire orchard (*viridarium*), De Sepibus concluded with some satisfaction that "your eye, being deceived by so excellent a labyrinth, thinks itself drawn forth into an infinite space and fields" (*oculus tuus tali labyrintho delusus in infinitum spacium, & campos protractum se putet*). But he also went on to say that "in physics the probable arguments of contrary opinions supply the daily material of disputation in the schools; here [in the museum] sight seems to establish conclusively by means of convincing appearance that infinity with respect to act is to be granted" (*in Physicis contrariarum sententiarum probabilia argumenta quotidianam in scholis subministrant disputandi materiam; hîc convincente visus apparentiâ conclusivè dari actu infinitum stabilire videtur*).

The language employed here by De Sepibus deserves further attention, but first it is worth noting that this juxtaposition between scholastic disputation and artificial demonstration draws an important parallel with the culture of "wonderful machines" described by Rivka Feldhay, who notes that "in Jesuit scientific culture of those days, public explanation of mathematical problems became a special kind of ritual, a display of knowledge close in spirit to the 'disputatio' and the public defense of theses."[35] Feldhay

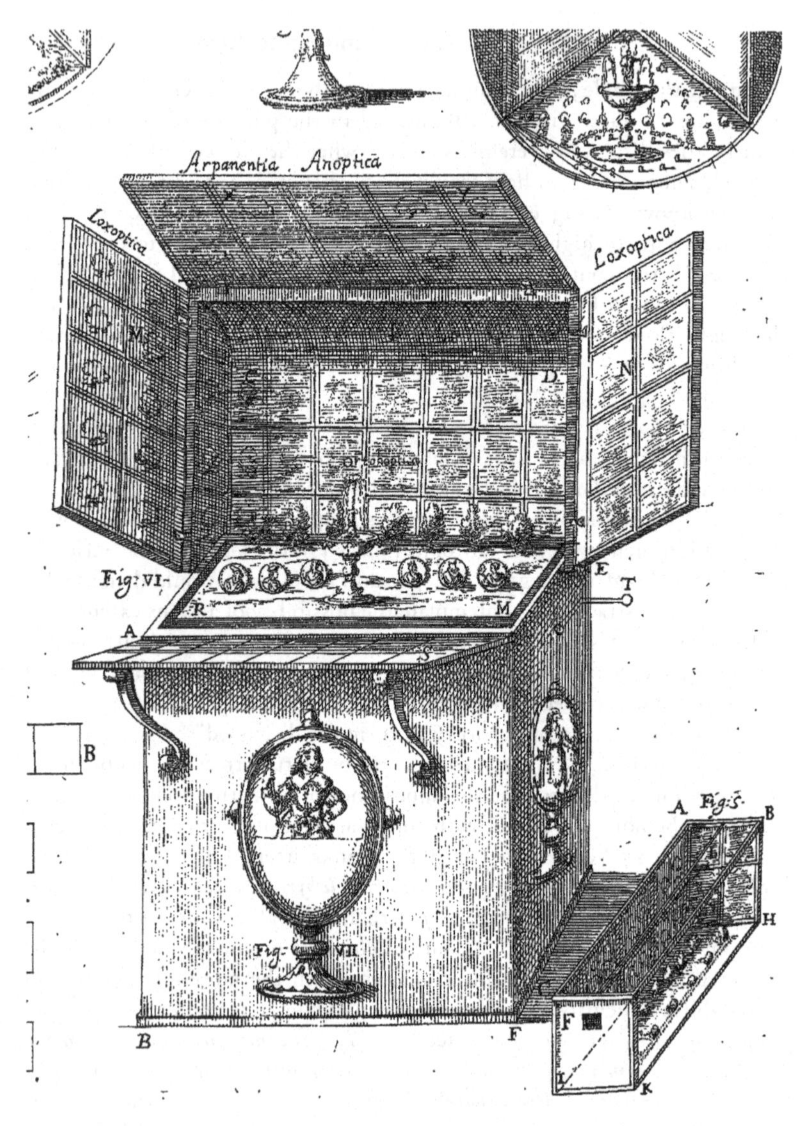

Figure 4.4 An illustration of the /theatrum catoptricum/, designed "so that reason is separated from the senses." From Kircher's /Ars Magna Lucis et Umbrae/ (1671 ed.). Courtesy of Department of Special Collections and University Archives, Stanford University Libraries.

focuses her attention on a Jesuit mathematician named Paolo Casati (1617–1707) who, in 1655, published his *Terra machinis mota* (The Earth Moved by Machines). Casati was active in the Society of Jesus at precisely the same time that Kircher's museum was becoming one of the intellectual centerpieces of the *Collegio Romano*, and the *theatrum catoptricum*, with its attention to the question of infinity, would seem to confirm that Kircher himself was involved in this culture of public explanation and display.

Note, too, how De Sepibus phrased his discussion of the *theatrum*. He created a clear juxtaposition between "the schools" on the one hand and the museum on the other, and between the "probable arguments" of the *disputatio* and the "conclusive" evidence provided by sight alone. But this was not a simple endorsement of empiricism, for De Sepibus was careful to note that the *theatrum seems* to establish conclusively only the possibility of infinite action, and that it does so with a "convincing appearance" of infinity. This was a reiteration of the same ambivalence toward the senses that permeated other parts of the Kircherian collection, though De Sepibus did suggest that even the optical trickery of the *theatrum* was better than the probable wrangling of the *disputatio*. What was important was not that the device in question could or could not create an action that was truly extended to infinity, but that with it Kircher could encourage the contemplation of such an action.

Further underscoring the seeming ambivalence toward the senses communicated by De Sepibus was the way in which he described how the *theatrum catoptricum* operated on the individual viewer. We know that he emphasized the ease with which the eye could be fooled "by so excellent a labyrinth," but he also considered what effect this would have on the mind. Kircher designed the *theatrum*, De Sepibus claimed, "so that reason is separated from the senses, or at least so that whomever you please is struck dumb by this instrument" (*ut rationem á sensu separatem, vel saltem deceptam in hoc instrumento quivis obstupescat*). Thus, to be "struck dumb" was to have one's reason separated from one's senses, to disrupt the orderly progression of impressions from the eyes to the mind.

In fact, I would argue that at least one purpose of this device and of the myriad other spectacular machines in the Kircherian collection was to engage in precisely this disruption of knowing, by driving a wedge between the senses and the mind. Such disruption is, perhaps, implicit in the very production of spectacle, which is intended to amaze and, indeed, to "strike dumb." Was this comparable to the self-conscious disconnect fostered by early modern skeptics between the (fallible) senses and the rational mind?[36] Perhaps, though Kircher was certainly not a "skeptic" in the sense of some of his contemporaries. Whatever messages concerning sensual fallibility his museum might have conveyed, it was at the same time an extravagant celebration of those same senses, with its painstaking and opulent attention to the fundamentals of baroque aesthetics. There was no disdain for the senses in the Kircherian collection, and certainly no lack of respect for their power to affect the mind. Nonetheless, these things were tempered by reminders that the senses could provide both pleasure and confusion. Ultimately, in its quiet commentary on the act of knowing, the Kircherian museum leavened the extravagance of baroque artistry with instruction that was both more sober and more subtle.

Significantly, the messages that Kircher encoded in these magnetic and catoptric machines also stood in sharp contrast to the basic tenets of scholastic psychology and epistemology. Surely, if De Sepibus and Kircher adhered even to a nominally Aristotelian way of knowing, we would expect them to uphold the scholastic maxim most famously elaborated by Thomas Aquinas: "Nothing exists in the mind that is not first in the senses." And yet, in the *theatrum catoptricum* and the other machines in the Kircherian museum, we find a playful but stubborn refusal to accommodate such an epistemology. Nor should this necessarily surprise us. Strict adherence to Aristotle (and to Aquinas) was an ideal rather than a universal reality in the Society, and a number of Jesuits were willing to adapt, or sidestep altogether, the more inconvenient tenets of their shared philosophy.

Of course, Kircher did not challenge the utility of the senses simply by leading his guests into a dark room, closing the door, and leaving it at that. There assuredly were things to observe and witness throughout the collection: even if one could not see the magnetic force that propelled Typhus across the ocean, one was treated to a lavish and beautiful spectacle of the effects of this propulsion. But if we take seriously Kircher's claims that his museum was a *theatrum mundi*, a reflection of the world, then we—like his baroque audiences—are left to grapple with a worldview in which much is obscured or hidden, revealed only by the playful use of analogy, correspondence, and the careful juxtaposition of Art and Nature.

In their own discussion of the mirrored machines that filled the Kircherian collection, Barbara Maria Stafford and Frances Terpak claim that "the mirror-lined cabinet of one sort or another was perhaps the prime embodiment of the Jesuit obsession with resemblance and synthesis, immutability and variation."[37] They describe the many-mirrored *theatrum polydictium* as an "analogical device [that] revealed how one entity could become pluralized into a multitudinous creation," much like the *theatrum catoptricum*, but they also highlight how central such machines could be to the Jesuit program of education and edification: "The complexities of visuality inform both Jesuit epistemology and their pedagogy.... In the hands of the Jesuits... such devices demonstrated the fickleness of the human mind, the duplicity of sensory appearances, and the convertibility of the physical universe."[38]

This analysis of such mirrored machines also calls attention to an important spiritual dimension, for "such conjuring devices were also spiritual tools for meditating on the knotty theological problem of divine absence and presence."[39] We have already seen that at least some of these devices were designed so as to permit meditations of a more secular nature on such abstracts as the concept of infinity. As Stafford and Terpak phrase it, "Jesuit catoptric experiments... can be thought of as mathematical, perspectival, and religious strategies for forcing hidden operations to manifest themselves."[40] Perhaps, then, these machines can be better understood as aids to a kind of secular meditation, permitting a repeated movement between the visible and invisible in the pursuit of understanding. In this, such technologies are again comparable to the magnetic devices discussed above, which encouraged the same movement between seen and unseen.

These mirrored cabinets and catoptric devices, then, were capable of conveying any number of messages to Kircher's audiences. Like his magnetic demonstrations, they were puzzles in need of a solution. But elsewhere in the museum, Kircher displayed machines that were less subtle in the messages they communicated. Consider, for example, a series of devices that appeared to work by means of perpetual motion, that dream of engineers and philosophers dating back to antiquity. In fact, these machines were little more than a clever trick designed to expose the ignorant and the unwary.

In his catalogue, De Sepibus emphasized that these devices displayed merely the *appearance* of perpetual motion (*De Mobili perpetuo apparente*). After offering his readers a quick refresher in Aristotle's definition of motion as the actualizing of potential and its division into "instantaneous and successive, violent and natural," De Sepibus observed that while in times past "a great many clever devices of philosophers and mathematicians exerted themselves" in the pursuit of perpetual motion, they had been "duped by a vain hope into wasting work and oil." Such devices existed only because these philosophers and mathematicians did not "understand the principles of nature."[41]

Having dismissed perpetual motion as a fantasy—and in doing so, upholding (at last!) a fundamental tenet of scholastic philosophy—De Sepibus then described a number of machines in the Kircherian collection that pretended to demonstrate precisely this kind of motion, first cautioning us to remember that these machines "do not profess perpetual motion, but [actually] refute it."[42] What is interesting about these devices is that, while they too were based on trickery and subterfuge (as were the magnetic and catoptrical machines), the goal in this case was *not* to be deceived. Their playful intention was to trick the spectator into believing that their motion truly was never-ending, but to believe this was to believe a philosophical fallacy. The magnetic and catoptrical machines illuminated even as they deceived, but these pretenses to perpetual motion did the opposite: to be taken in by the deception was to fail to understand "the principles of nature." Moreover, I suspect that one was made privy to the deception only by the intervention of Kircher himself, thereby echoing the principles at work in the magnetic machines whose mysterious innards and mechanisms were made obvious only by their deliberate revelation.

A Kircherian Epistemology?

Michael Gorman has asked, "What . . . if Kircher never had any intention of creating certain and demonstrable knowledge? What if his more humble goal was to accumulate and disseminate a body of probable knowledge?"[43] He notes, too, that such questions are rarely, if ever asked of Kircher, who is instead subjected by historians to "alien epistemological standards" of which, inevitably, he falls short. Gorman goes on to link Kircher's epistemology with the senses, though not in a fashion consonant with standard scholasticism:

> Kircher's presentation of himself as a mediator of the opinions and observations of others rather than a forger of new dogmas and certainties was... a position that was well adapted to the intellectual climate of Rome... [in] a period characterized by increasingly fervent persecution of Jesuits who deviated from Aristotelian orthodoxy in matters of natural philosophy. Kircher's [ideal] observer... embodies a powerful epistemological stance, at the center of which lies individual sensory weakness and fallibility.[44]

While I disagree with Gorman's insistence on painting Kircher as a "mediator" rather than as an active proponent of a particular way of studying nature, his overall point here remains an astute one. He links Kircher's refusal to present demonstrative certainty with his attention to the "sensory weakness and fallibility" that was undeniably demonstrated by the machines and displays scattered throughout his museum.

In a similar vein, Stuart Clark has characterized Kircher's "wonderful works of light and shadow" as an "attempt to unsettle 'the grammar of visibility,'" adding that

> the spectator is made to feel—besides bemusement, presumably—that what is seen is not determined by anything that is fixed in the visual environment but depends instead on his or her own interpretation and viewing point. Perspectival optics has become fluid and actively creative, as well as an indicator of the insubstantiality of appearances, which are "flaunted" to the point of "epistemological unreliability."[45]

A word like "unsettle" seems appropriate to the Kircherian collection—much like the notion of separating the senses from the mind, or even the more general ideal of bemusement, it implies that Kircher set out deliberately to undermine the very foundations of knowing, and thereby the epistemological confidence of his audiences. As Clark puts it, there was an unmistakable element in these machines of "epistemological unreliability."

I suggested above that the machines in the museum trained their audiences in *how to look*. They accomplished this, in part, by unsettling the grammar of visibility, to borrow Clark's phrase. In almost every case these devices embodied a critique of the senses, with their emphasis on *seeming* and *appearing* or, conversely, on *hiding* and *obscuring*. Unique as they were to this collection, and forming as they do the core of its catalogue, we can reasonably assume that these machines represented a central expression of a "Kircherian" epistemology.

By creating and exaggerating this "epistemological unreliability" throughout his collection, Kircher was able simultaneously to erode the foundations of one world—as perceived through the lens of conventional scholastic natural philosophy—and replace them with his own vision of the cosmos. The world that Kircher constructed was balanced upon a series of unseen and mysterious foundations, its disparate parts linked by the "secret knots" that Kircher referenced in his works on magnetism. Thus, in its identity as a *theatrum mundi*, the Kircherian collection was fundamentally a theater of the unseen in which the secrets of Nature were simultaneously displayed and controlled by Kircher's elaborate, playful machines. Moreover, it was through those machines that Kircher's visitors were first introduced to this mysterious new world, and then educated both in the limits of their understanding and in the tools they might use to push those limits further still.

Such a world may well have been unsettling to baroque audiences, cloaked as it was in metaphorical shadows that resisted conventional observation. Balancing this, however, was Kircher's indefatigable optimism. Where contemporary skeptics despaired of ever knowing Nature in its entirety, Kircher remained convinced that understanding was not only possible, but also inevitable. His unswerving faith in a providential God made him certain that, no matter how shadowed and hidden the foundations of the world, humanity could and would come to see them. The Kircherian museum, with its elaborate machines and its repeated exhortations to see with the eyes of the mind as well as those of the body, embodied the means whereby, in Kircher's own words, "the hidden is led from the most profound shadows into the astonishing light."[46]

Notes

1. Paula Findlen, "Scientific Spectacle in Baroque Rome: Athanasius Kircher and the Roman College Museum," in Mordechai Feingold, ed. *Jesuit Science and the Republic of Letters* (Cambridge, MA: MIT Press, 2003), 225.
2. For more on Clavius and Grienberger, see Ugo Baldini, "The Academy of Mathematics of the Collegio Romano from 1553 to 1612," in Feingold, *Jesuit Science and the Republic of Letters.*
3. Michael John Gorman, "From 'The Eyes of All' to 'Useful Quarries in Philosophy and Good Literature': Consuming Jesuit Science, 1600–1665," in John W. O'Malley, S.J., Gauvin Alexander Bailey, Steven J. Harris, and T. Frank Kennedy, S.J., eds. *The Jesuits: Cultures, Sciences, and the Arts, 1540–1773* (Toronto: University of Toronto Press, 1999), 179.
4. Findlen, "Scientific Spectacle," 240.
5. Paula Findlen, *Possessing Nature: Museums, Collecting, and Scientific Culture in Early Modern Italy* (Berkeley: University of California Press, 1994), 84.

6. Findlen, "Scientific Spectacle," 229.
7. Gorman, "From 'The Eyes of All,'" 171.
8. Gorman himself suggests a connection with the ideal of *sprezzatura*: see his "Between the Demonic and the Miraculous: Athanasius Kircher and the Baroque Culture of Machines," in Daniel Stolzenberg, ed. *The Great Art of Knowing: The Baroque Encyclopedia of Athanasius Kircher* (Stanford: Stanford University Libraries, 2001), 62.
9. Giorgio de Sepibus, *Romanii Collegii Societatis Jesu Musaeum Celeberrimum* (Amsterdam: Ex Officina Janssonio-Waesbergiana, 1678), *Praefatio ad Curioso Lectori*: "Artis itaque & Naturae ergasterium, disciplinarum Mathematicum gazophylacium, philosophiae practicae epitomen, Musaeum Kircherianum hisce..."
10. Findlen, *Possessing Nature*, 85.
11. This trend is described in excellent detail by Findlen in her *Possessing Nature.*
12. Peter Dear, *Discipline and Experience: The Mathematical Way in the Scientific Revolution* (Chicago: University of Chicago Press, 1995), 43–46.
13. Janet Vertesi describes some of these strategies in her "Picturing the Moon: Hevelius's and Riccioli's Visual Debate," *Studies in History and Philosophy of Science Part A* 38 (2007): 401–421.
14. Dear, *Discipline and Experience,* 19.
15. A. Lugli, "Inquiry as Collection: The Athanasius Kircher Museum in Rome," *Res* 12 (Autumn 1986): 109–124.
16. Findlen, *Possessing Nature*, 93.
17. De Sepibus, *Romanii Collegii Societatis Jesu Musaeum Celeberrimum,* 2.
18. Ibid., 30: "Habet *Musaeum* tria *Rhinocerotum Cornua*..."
19. Ibid., 25
20. Ibid., 1: "Ad dexteram latus Bibliothecium solius authoris operibus refertum conspicitur...Ad sinistrum latus pulpitum versatile 12 Romanorum Imperatorum copiosis numismatibus pretiosum...Quod sequitur lapidariae artis, & sculptoriae refertissima mensa...."
21. Ibid., 18: "Magnes subtilium cos ingeniorum, prodigium Naturae, mirabilium thaumaturgus, reconditae virtutis inscrutabilis Labyrinthus, *Kircheriani Musæi* medulla, lapis est...."
22. Ibid., 19.
23. Athanasius Kircher, *Magnes; sive, De arte magnetica,* 2nd ed. (1643), 319. On Daedalus's animation of a statue of Aphrodite, see William R. Newman, *Promethean Ambitions: Alchemy and the Quest to Perfect Nature* (Chicago: University of Chicago Press, 2004), 12–13.
24. On the rise of the imagination in early modern culture and its implications for philosophy and psychology, see Stuart Clark, *Vanities of the Eye: Vision in Early Modern European Culture* (Oxford: Oxford University Press, 2007), esp. 42–45.
25. For more on this, see David Freedberg, *The Power of Images: Studies in the History and Theory of Response* (Chicago: University of Chicago Press, 1989), 160–185.

26. Perhaps the most striking and beautiful examples of this meditative literature can be found in two works commissioned by Geronimo Nadal (1507–1580), the *Evangelicae historiae imagines* of 1593 and the expanded *Adnotationes et meditationes in Evangelia* of 1595. See Walter S. Melion, "Memory, Place, and Mission in Hieronymus Natalis' *Evangelicae historiae imagines*," in *Memory and Oblivion: Proceedings of the Congress of the History of Art held in Amsterdam, 1--7 September 1996* (Dordrecht: Kluwer Academic, 1999), 603–608.
27. Mark A. Waddell, "The World, As It Might Be: Iconography and Probabilism in the *Mundus subterraneus* of Athanasius Kircher," *Centaurus* 48, no. 1 (2006): 3–23.
28. See, for example, Judi Loach, "The Teaching of Emblematics and Other Symbolic Imagery by Jesuits within Town Colleges in 17th- and 18th-Century France," in John Manning and Marc van Vaec, eds. *The Jesuits and the Emblem Tradition: Selected Papers of the Leuven International Emblem Conference, 18–23 August, 1996* (Turnhout, Belgium: Brepols, 1999), 161–186. See also G. Richard Dimler, S.J., "Humanism and the Rise of the Jesuit Emblem," in Peter M. Daly, ed. *Emblematic Perceptions: Essays in Honor of William S. Heckscher* (Baden-Baden: Verlag Valentin Koerner, 1997), 98–101 in particular.
29. Brian S. Baigrie, "Descartes's Scientific Illustrations and 'la grande mécanique de la nature,'" in Baigrie, ed. *Picturing Knowledge: Historical and Philosophical Problems Concerning the Use of Art in Science* (Toronto: University of Toronto Press, 1996), 111.
30. Ibid., 123.
31. Stephen Menn, *Descartes and Augustine* (Cambridge: Cambridge University Press, 1998), 215.
32. A number of historians have noted the strong links between similitude and the imagination; as one example, see Katharine Park, Lorraine Daston, and Peter L. Galison, "Bacon, Galileo and Descartes on Imagination and Analogy," *Isis* 75, no. 2 (1984): 287–289, as well as the essays that follow this introduction.
33. De Sepibus, *Romanii Collegii Societatis Jesu Musaeum Celeberrimum,* 36: "*Theatrum Catoptricum*, instrumentum est, quo quæstionem inter Physicos tam celebrem *de Infinito actu*..."
34. As one example of the long history of this question, as well as its theological contexts, see Anne A. Davenport, "The Catholics, the Cathars, and the Concept of Infinity in the Thirteenth Century," *Isis* 88, no. 2 (June 1997): 263–295.
35. Rivka Feldhay, "On Wonderful Machines: The Transmission of Mechanical Knowledge by Jesuits," *Science and Education* 15 (2006): 153.
36. For more on this, see Clark's *Vanities of the Eye*, esp. Ch. 8.
37. Barbara Maria Stafford and Frances Terpak, *Devices of Wonder: From the World in a Box to Images on a Screen* (Los Angeles: Getty, 2001), 26.
38. Ibid., 27.
39. Ibid.

40. Ibid., 28.
41. De Sepibus, *Romanii Collegii Societatis Jesu Musaeum Celeberrimum,* 56: "...his praepositis de possibilitate dabilis motus perpetui inquiritur, in quo tot jam saeculis plurima Philosophorum, Mathematicorumque ingenia desudarunt, & dum operam oleumque perdidere vana spe delusi, se naturae principia non intelligere...."
42. Ibid.
43. Michael John Gorman, "The Angel and the Compass: Athanasius Kircher's Magnetic Geography," in Findlen, ed. *Athanasius Kircher: The Last Man Who Knew Everything,* 255.
44. Ibid., 255–256.
45. Clark, *Vanities of the Eye,* 101.
46. Athanasius Kircher, S.J., *Ars magna lucis et umbrae, in decem libros digesta* (Rome: Hermanni Scheus, 1646), 834: "Haec autem est divina illa Optice scientia, quae quod abditum est è profundissimis tenebris in admirabile lumen educit."

CHAPTER 5

Fantasy Islands: *Utopia, The Tempest,* and *New Atlantis* as Places of Controlled Credulousness

Guido Giglioni

Historians have long acknowledged a number of intriguing family resemblances between Thomas More's *Utopia*, William Shakespeare's *The Tempest,* and Francis Bacon's *New Atlantis*. Allusions, cross-references, and intertextual connections have been duly pointed out and explored.[1] In this essay I focus on the role that suspended disbelief plays in the worlds imagined in the three works, for Utopia, Prospero's island, and Bensalem share the characteristic of being islands of controlled credulousness. By focusing on *persuaded, bewitched,* and *manipulated* imaginations in *Utopia*, *The Tempest,* and *New Atlantis*, I investigate the nature of the relationship that exists between what can be imagined and what can be believed, and how this relationship affects the world structures in which one can operate. Enchantments, be they literal or metaphorical (but is this distinction really applicable?), could be used to obtain magical, aesthetic, or political forms of consensus. In different ways, the three works deal with the magic of words. In this, they reflect a debate that was common at the time. Magic became a particularly controversial issue in the early modern period because it represented an *oblique*, *rough,* and *secret* way of gaining consensus, difficult to control, both for the practitioner and the client, the operator and the victim, the enchanter and the enchanted.

My use of the adjectives "oblique," "rough," and "secret" is not accidental. They signal strategies of control that play a crucial role in the works under examination. *Oblique* is for More the only type of philosophical discourse that may find a political use in human society. *Rough* is the type of magic that, by engaging with supernatural forces, allows Prospero to produce effects that go beyond the level of mere illusion. *Secret*, finally, is the type of knowledge that, by circumventing conscious intentions, manages to govern an otherwise riotous commonwealth of appetites. Some of the basic questions that one can find addressed in *Utopia*, *The Tempest*, and *New Atlantis* can, therefore, be described as attempts to qualify the relationship between patterns of persuasion and man's disposition to believe: How can the work of the imagination become an object of belief? What ends and means are considered legitimate to gain persuasion and transform imaginable into believable objects? What kind of influence are we entitled to exercise upon other human beings in order to bring about political change, social betterment, and ethical regeneration, or, more simply put, to rebuild the world according to our imaginations?

The early modern tension between imagination and belief was symptomatic of a condition of diffuse anxiety concerning the limits of human credulousness and the nature of the equivocal space between reality and appearance, reason and delusion, literal and figurative representations of things. In a recent study, Stuart Clark demonstrated how controversial visual experience had become between the sixteenth and the seventeenth century. In *Vanities of the Eye*, he meticulously reconstructed the cultural circumstances that led to a generalized mistrust of appearances and semblances. One has only to think of the wealth of demonological literature produced during that period, the study of perspective and anamorphic art, the early modern revival of ancient skepticism, the rejection of all forms of idolatrous worship and questioning of the very visibility of spiritual reality by Protestant reformers, to mention only some of the topics examined by Clark.[2] Building upon such a framework, this essay intends to concentrate on the mimetic, empathetic, and projective powers of the imagination in three different literary contexts and to investigate the nature of virtual reality created by the imagination in states of visual and cognitive irresolution between belief and disbelief.

Islands of Subverted Reality

As a result of the Renaissance discovery of new worlds, islands became *topoi* of philosophical debate. They provided the ideal background against which writers could introduce new political ideas, discuss the anthropological, ethnographic, and religious consequences of expanded geographical horizons, and attempt subversions of hierarchical relations, critiques of conventional wisdom, and experiments in counterfactual arguing. Finally, through the figure of the island, they could symbolize such philosophical notions as individuality, isolation, totality, and self-sufficiency. In this, Utopia, Prospero's island, and Bensalem are typical products of what Roland Greene has called "island logic."[3]

Being an island means to be separated from, and yet inevitably defined by, the mainland. Notions of difference and discontinuity are an integral part of every island's essence. A trait that, according to More, reveals the Utopians' strength and uniqueness is that originally their country was not an island, but became so when the isthmus connecting the peninsula to the continent was cut away through human labor. As a result of their tenacious and collective work, Utopians managed to distinguish themselves from the rest of humankind: "the neighbouring people, who at first had laughed at the folly of the undertaking, were struck with wonder (*admiratio*) and terror (*terror*) at its success."[4] By cutting their ties with the continent, Utopians sanctioned their special status as a self-existing, autarkic community.

In terms of narrative devices, then, it is certainly no accident that castaways are likely to discover a new island after they survive a shipwreck. The spatial isolation of the island corresponds to the abrupt temporal discontinuity caused by the shipwreck. *Utopia*, *The Tempest,* and *New Atlantis* are as much about islands as they are about shipwrecks, although no *real*, literal shipwreck takes place in any of the three works. In *The Tempest* the shipwreck that opens the play is staged; in *New Atlantis,* the shipwreck (better, the mysterious, forced landing) is in all likelihood engineered by the technologically savvy Fathers of Solomon's House; in *Utopia*, finally, the shipwreck is figurative, namely, the "shipwreck" of Western Christianity and English society, both ravaged by wars, social disruption, and economic crisis. To be fair, one shipwreck is actually recorded in *Utopia*: "some twelve hundred years ago," recounts Raphael Hythloday, "a ship which a storm had blown towards Utopia was wrecked on their island. Some Romans and Egyptians were cast ashore, and never departed."[5] Bacon expands on the importance of accidental landings in his *New Atlantis*. Only rarely and very reluctantly do foreigners who happen to land on Bensalem decide to leave the island. It is part of the "island logic" to become insular, make a totality out of a single instance, grow as a self-sufficient and organic unit, and sever the ties with the rest of the existing reality.

All this means that a writer, in dealing with shipwrecks and islands, can rely on material that is extraordinarily malleable for the imagination. More, Shakespeare and Bacon play with varying degrees of fictional reality. In *Utopia*, the relationship between reality and fiction is very complex and subtle. Hythloday contrasts English laws and customs with laws and customs of imaginary peoples: Utopians, of course, but also Polylerites ("much nonsense"), Achorians ("without place"), Macarians ("the blessed"), Anemolians ("windy"), and Zapoletes ("busy sellers"). The more outlandish the level of fictional invention is, the more acute the sense of reality becomes. This demands urgent political action. It is precisely through "much nonsense" that More can display a no-nonsense attitude in his analysis of contemporary English reality. "*Finge me apud regem esse Gallorum*" (imagine that I am at the court of the King of France) says Hythloday in the middle of his discussion on Plato's ideas about politics.[6] The description of an imaginary royal council represents for More the opportunity to provide a sample of Machiavellian politics. *Utopia* can be seen as a series of experiments in political imagination, in which philosophical paradoxes, fictional devices, and exercises in religious exegesis are used to debate possible solutions to policy matters. More sketches the boundaries of an imaginary geography to stimulate the reader's imagination and his or her freedom to contrast the real with the ideal.[7]

The very word "utopia" signals the Platonic paradox that "real" reality (όντως ὂν) is not something to which one could have access in this world. In *The Tempest*, Shakespeare hints at real geographical places, but in fact the whole landscape is the imaginary setting of a romance. Like Shakespeare, Bacon refers to actual news of contemporary shipwrecks, but the real event is only the pretext for forays into the domain of virtual worlds. Bacon's view of alternative worlds, though, lacks the sense of intelligible transparency that underlies More's *Utopia*, where the tension toward ideal patterns of rationality clearly indicates that the legacy of Plato's idealism is still powerful. *New Atlantis*, on the contrary, shares with *The Tempest* an atmosphere of eerie opaqueness that may be said to belong to the rising genre of science fiction more than utopian literature.[8] The sense of precarious poise that surrounds the three islands due to a skillful mixing of real events, idealized reality, and suspended imagination is reinforced by the use of ironic, ambivalent, and tragicomic effects. This aspect is particularly relevant to this essay, for the sense of suspended tension between reality and appearance is constantly heightened by stylistic devices that deliberately subvert established relationships between truth and its representation.[9]

Such effects of subverted reality result from different writing strategies in the three works. The tragicomic tone of *Utopia* is due to the contrast between reality and ideality, so that the playful characterization of imaginary societies accentuates the tragic character of fifteenth-century English social and political reality.[10] In *The Tempest*, the sense that events unfold in a tragicomic fashion results mainly from the contrast between illusion and delusion.

Finally, if Bacon's *New Atlantis* may seem to resonate less with tragicomic accents, this is because the interplay of seriousness and playfulness is created by a dramatic situation in which different groups of people act according to different levels of knowledge that are not always shared by all the participants in the action. The sharp contrast between the state of ignorance in which the castaways are kept by the Fathers of the Solomon House and the strikingly high level of knowledge underlying Bensalemite technology inevitably produces a spooky atmosphere of dark humor or tragicomic cunning of reason. Whether the contrast is between reality and ideality, illusion and delusion, openness and secrecy, in all three cases the result is a sense of tragicomic imbalance.[11]

Ironic refractions and subverted perspectives are reinforced by stylistic devices and choices of literary genres. *Utopia* is a dialogue within a dialogue, *The Tempest* contains plays within the play, and *New Atlantis* is the story of a gradual revelation as is retold by a narrator through the use of reported speeches. Admittedly, *New Atlantis*, unlike the dialogical *Utopia* and the dramatic *Tempest*, seems to contain a more unitary point of view corresponding to the anonymous narrator of the story. However, that identity remains nonetheless blurred due to the very anonymity of the narrator and because in telling the story the narrator oscillates between the singular ("I") and the plural personal pronoun (the "we" of the crew). On the one hand, the progression toward an increasingly focused point of view reflects the anonymous narrator's gradual raising in awareness; on the other hand, the more the narrator becomes acquainted with the mechanisms that govern the organization of Bensalem, the more the reader is overwhelmed by the opaque character of the whole enterprise. In *The Tempest*, the question of the point of view is even more complicated, for the voices are constitutively multiple from the very beginning. The dramatic structure, though, does not prevent Prospero from acting as a central focus: in the drama within the drama, he is the stage director, while the rest of the characters are actors who play the role that he has assigned them. More, finally, is recounting his story in the first person—the king of England "sent me into Flanders as his spokesman"—and yet even his point of view shifts constantly from narrator to character.[12] While in *New Atlantis* the narrator is clearly different from the author, in *Utopia* the author and the first-person narrator seem always to coalesce.

Finally, it is the very genre of the three works that blurs the boundaries between faction and fiction, between truth and its ironic subversion. *Utopia* bears the mark of the playful and disillusioned critique of mores and traditions that is the defining character of the humanist inquiry; *The Tempest* takes advantage of the state of willing credulousness created by theatrical staging; *New Atlantis* conveys the feeling of disorientation that derives from traveling to foreign, unknown lands and that is the distinctive trait of travel literature. In different ways, the narrative strategies exploit the mimetic, projective, and empathetic tendencies of the reader. An important role in this process of literary make-believe is played by the three master-tellers: Raphael Hythloday in *Utopia*, Prospero in *The Tempest,* and the "anonymous narrator" in *New Atlantis*. To qualify the three master tale-tellers and their modus operandi in a more concrete way, I introduce three "figures" of authorial delivery: the "midwife," the "conjuror," and the "spy." They correspond respectively to Hythloday's Socratic impersonation, Prospero's exercise of "rough" magic and, finally, the law of "knowing-without-being-known" implemented by the Fathers of Solomon House in Bensalem. Ironically enough, for all their differences, the *midwife*, the *conjuror,* and the *spy* share the quality of being barren of knowledge of their own: the

Socratic midwife facilitates other people's recollection, the conjuror makes things appear or occur by using intermediate agents, the spy collects intelligence from other people's knowledge.

Believable Imaginations

There are situations in which people believe that what they are imagining is real. A person who is having a dream during sleep, for instance, has often no awareness of being dreaming. The same may happen in states of intense day-dreaming, visions, and hallucinations. Besides these cognitive situations in which beliefs and imaginations seem to blend together effortlessly, there are less seamless and cogent experiences of believing the imaginable, in which the knowing subject oscillates between willingness and resistance, awareness and oblivion, surrendering in the end to a temporary and precarious acceptance of his or her perceived reality. After Coleridge's celebrated essay, the technical term for such a condition is suspension of disbelief.[13] A notoriously complex case of suspended disbelief (a suspension in which the level of coercion is higher than the one required in theatrical representations and religious rituals) is a magical enchantment. Casting spells may be seen as a way of taking advantage of the supple nature of one's own and other people's imagination in order to impose beliefs.

The special relationship between imagination and belief is a time-honored topic in the history of both philosophy and medicine, not to mention, of course, the field of aesthetic theory. Here I will confine myself to some cursory remarks to introduce the theme of suspended disbelief in *Utopia*, *The Tempest,* and *New Atlantis*. While Plato considered the imagination to be yet another variety of belief (*Theaetetus*, 152C–160C; 164D–165B), Aristotle entrusted the same faculty with a specific and autonomous cognitive domain, characterized, unlike belief, by the ability to recall images at pleasure (we cannot believe at pleasure) and by the ability to partially distance ourselves from the content of our imagination so as to be like spectators in a theater (again, we cannot distance ourselves with the same ease from our beliefs (*De anima*, III, iii, 427b). Aristotle thus described the imagination as a cognitive state in which the knowing subject is able to suspend his or her credulousness, a characteristic that he did not allow to belief.[14]

To this picture we should add the far-reaching contribution of the medical tradition. Physicians throughout the centuries tended to identify imagination and belief as one vital, healing power (the renowned *virtus medicatrix naturae*), so that belief (also characterized as *fides* or *confidentia*) was deemed to give direction and structure to the transformative faculty of the imagination, while the imagination could amplify the changeable nature of the body. The conclusion was that belief and imagination were capable of reinforcing each other in numerous ways.[15] It is also worth remembering that the "suspendible" nature of the imagination has always attracted the attention of philosophers interested in providing therapies of desire. Epicureans, Stoics, and Sceptics, for instance, despite the notable differences in their targets (*atarasseia*, *apatheia*, and *epoche*), advocated a more radical approach: they thought it possible to convert the suspension of disbelief provided by the imagination into a state of fully suspended belief.[16]

In their treatment of believable imaginations, suspended judgments and overpowering spells, More, Bacon, and Shakespeare refer to well-known patterns of argumentation and persuasion. In different ways, they all describe situations in which the power of the imagination, with its bodily energy and infinite combinatorial possibilities, coalesces with the sense of assurance offered by the act of believing. In Bacon's case, the imagination coincides with the innermost vitality of nature and matter, together with their appetites and ultimate tendencies. He is well aware that the imagination has a great power over material nature. He tends to believe that a large part of the "facts" and "stories" concerning the extraordinary effects (wonders) of the imagination are real and reliable. What is more, he is eager to find ways of harnessing this source of vital energy diffused in nature and to apply it to concrete situations of both natural and human life. However, he is also perfectly aware that a fundamental discrepancy separates the sphere of belief from that of the imagination. In this he remains an Aristotelian. "I cannot command myself to believe what I will," he states in *Sylva Sylvarum*.[17]

The same situation is acknowledged by More when he examines the religious customs of the Utopians. They Utopians are in favor of religious tolerance because "they are persuaded that no one can choose to believe by a mere act of the will."[18] While they reject magical beliefs and divinatory practices as vain, they understand the social relevance of religious beliefs: "they venerate miracles which occur without the help of nature, considering them direct and visible manifestations of the divinity. Sometimes in great and dangerous crises they pray publicly for a miracle, which they then anticipate with great confidence, and obtain."[19] Although the practice looks like any other ritual of wonder-working, in More's opinion, Utopians' faith in miracle is not magic, but a pious practice meant to obtain divine favors through prayers. Agrippa of Nettesheim, another humanist who, like More, liked to engage in playful seriousness, was not contrary to such a use of prayers based on the magic of words.[20] What is more, More does not go so far as to assign the imagination, seen as a natural faculty, the power to induce material effects in the macrocosm and the microcosm. Being closer to Plato's view, More maintains that human imaginations are just like other forms of belief, only more effete and less reliable. Bacon, on the other hand, being convinced like Aristotle that belief and imagination belong to different faculties of the soul, is keen to find a way of bridging the gap between the two. He is clearly determined to exploit the natural process whereby what can be imagined is made believable. This is an experimental result, as it were, which is all the more urgent because, in Bacon's opinion, people need to believe before they can do or be told to do things.

Bacon provides a complex but illuminating discussion of the question of the believability of the imagination in the Experiments 945 and 946, Century 10, of his *Sylva Sylvarum*. Here he defines the imagination as "the representation of an individual thought" that has the power to move back and forth through time. The imagination is an essential component of man's thinking activity, which allows one to think of the future (belief), the past (memory), and the present (by visualizing things *as if* they were present, which is the highest degree of imaginative freedom, "imaginations feigned and at pleasure").[21] Bacon is particularly concerned with the question of the extent to which we can believe what we imagine, more specifically "whether a man constantly and strongly believing that such a thing shall be, (as that such an one will love him, or that such an one shall recover a sickness, or the like,) it doth help any thing to the effecting of the thing itself."[22] Bacon does not hesitate to make clear that here he is not talking about the effects that conscious beliefs and imaginations have on one's mind and body, as when one becomes more industrious by concentrating on the very idea of industriousness and by convincing oneself of the importance of being industrious. Rather, he is interested in those cases when one is not aware that the imagination has the power of shaping his or her beliefs "merely by a secret operation, or binding, or changing the spirit of another." Bacon, however, recognizes that the question is extremely difficult to approach because the investigator is confronted with two main experimental obstacles: first, beliefs cannot be self-induced ("I cannot command myself to believe what I will"); second, fear (or other intense emotional responses of a negative nature) is needed to make imaginations effective ("for a man representeth that oftener that he feareth"). The consequence is that such experiments of self-induced beliefs and imagination can be painful and dangerous for one's mental stability.[23] The only way of submitting the elusive phenomenon of the imagination to experimental and controllable conditions is by splitting the imagining and believing subject into two individuals, that is, a believer (or "actor") and an imagining subject (whom Bacon calls an "imaginant" or "imaginer") and by inducing beliefs in the "believer" through the "imaginant."[24] Bacon's "imaginant" functions as a medium, a sort of imaginative machine that can be used and manipulated in a controllable environment for the ends of the operator:

> The help therefore is, for a man to work by another, in whom he may create belief, and not by himself; until himself have found by experience, that imagination doth prevail; for then experience worketh in himself belief; if the belief that such a thing shall be, be joined with a belief that his imagination may procure it.[25]

One such experiment of disassociation devised in order to disentangle the components involved in the phenomenon of believable imagination did in fact take place in Bacon's own house. In the same passage of *Sylva Sylvarum*, he recounts his meeting with a juggler who boasted to be able to tell a card picked by someone (without seeing it), through an exercise of mind reading. Bacon explains that the juggler was not able to tell directly the card thought by the man because no one apart from God can have access to someone else's thoughts: one can only induce thoughts in someone else. Therefore, the juggler asked a second person to tell the card thought by the first person. The reason is that the first person (the "imaginant"), being in awe of the alleged powers of the juggler, had the ability to develop a strong imagination and influence the second person ("the believer" or "actor"), which is something that the juggler could not do because his imagination was too disbelieving. Instead, by developing a strong imagination, the "imaginant" could influence—"bind," to use the technical term—the thinking activity of the "believer." Bacon presents two scenarios and two possible interpretations of the phenomenon: either the juggler told the first man to think a card and then whispered in the ear of the second man what card the first man had thought or the juggler whispered in the ear of the first man the card he should think and then asked the second man what card the first man had thought. In this situation, the first man acted as the "imaginant," capable of shaping the "believer"'s thoughts. While the first scenario is in principle impossible because it violates a fundamental law of nature and a theological dogma—nobody can read other people's thoughts—the second scenario is the proof that thoughts and beliefs need the energy of the imagination to become effective.[26] The experimental situation of the juggler and the two servants described in *Sylva sylvarum* finds a narrative and fictional counterpart on a larger scale in *New Atlantis*. In the island, it is the whole people of Bensalem who release the operative energy of their imagination every time their beliefs are dramatically manipulated by the Fathers of Solomon's House, who have laboratories and workshops through which they can produce appearances of every kind. This makes the "enchanting" of the imagination possible from a technological point of view and on a collective level.[27]

Bacon's description of the procedure to induce beliefs in other people by stimulating their imaginations presupposes the operator's ability to arouse affects and create auras of authority and credibility. This point would not have convinced More, who was reluctant to admit that human imaginations could be manipulated into beliefs. The only aspect of the imagination in which he is interested is the power of devising virtual worlds through philosophical arguments.[28] The fact remains that Plato's, and thus More's, philosophical imagination, with its characteristic attitude of serious jest, is different from the condition of suspended disbelief that, according to Aristotle, remains the defining trait of the imagination. As for the place of believable imagination in Prospero's island, the imagination, besides being open to the intervention of supernatural forces (Prospero's magic is "rough"), has a remarkable fictive energy of its own that makes it liable to all sorts of illusions. *The Tempest*'s bewitched island is soaked with imagination to such an extent that Gonzalo bursts into the exclamation: "[a]ll torment, trouble, wonder and amazement / Inhabits here" (5.1.104–105). The power to induce beliefs demonstrated by Shakespeare's Prospero rests on the radical assumption that both beliefs and disbeliefs can be suspended, which creates a pervasive sense of sceptical *epoche* and cognitive uncertainty throughout the play—what Prospero calls the "subtleties o' th'isle," which "will not let you / Believe things certain" (5.1.124–125). More maintains a skeptical stance toward the imagination, deemed to be good for devising alternative hypotheses on reality, but not for changing reality. Shakespeare in turn maintains a skeptical stance toward reason, which is presented as unable to contrast the overpowering force of the imagination. In Bacon's case, the imagination can alter both reality and beliefs about reality. However, the fictive energy of the mind is not something of which human beings should be proud, for, writes Bacon in *Historia naturalis et experimentalis*, "we copy the sin of our first parents while we suffer for it." And he continues:

> They wished to be like God, but their posterity wish to be even greater. For we create worlds, we direct and domineer over nature, we will have it that all things are as in our folly we think they should be, not as seems fittest to Divine wisdom, or as they are found to be in fact; and I know not whether we more distort the facts of nature or our own wits; but we clearly impress the stamp of our own image on the creatures and works of God, instead of carefully examining and recognizing in them the stamp of the Creator himself.[29]

By creating fictive worlds, human beings reenact the original sin of wishing to be like God. Indeed, their sin is even greater, for the protoplasts wanted to emulate their Creator, while postlapsarian humans are convinced that they can create new reality without help from God. They believe the figments of their imagination to be more real than reality itself and more rational than God's norms of intelligibility. The sin lies in the view that "all things are as in our folly we think they should be" so that we impose our "image" on reality rather than discover the traces of God's wisdom in nature.

Another reason behind Bacon's attractions toward and cautions about the imagination's power to shape human beliefs and to suspend the mind's critical awareness of reality is that, by its very nature, the mind is drawn to focus and concentration, and avoids states of suspense (*ut non pensilis sit*). As Bacon argues in *De augmentis scientiarum*, "as in all motion there is some point quiescent"—the Atlas of the ancient fables—"so do men earnestly desire to have within them an Atlas or axletree of the thoughts (*cogitationum Atlas aut poli*), by which the fluctuations and dizziness of the understanding (*intellectus fluctuationes et vertigines*) may be to some extent controlled."[30] Suspense, though, is precisely the experience that, on the three islands, defines the tension between identification and estrangement, between imagination and disbelief.

Varieties of Suspended Imagination

Utopia: *The Unbelievability of Philosophical Reason*

Delusion (to believe that reality is the way we imagine it to be) and suspension of critical judgment (to concede for a little while, in a state of half-awareness, that reality is the way we imagine it to be) are aspects of the imagination that, in More's opinion, are sufficiently dangerous for the human mind's sanity to require careful consideration. The specific problem that More (the fictional More) addresses to Hythloday's apolitical stance is that philosophers have become "precious": they "do not condescend even to assist kings with their counsels."[31] In an important passage during their conversation, Hythloday tells More (the character) that, even if he decided to divulge his political ideas, tested by experience based on countless travels and long meditations, people would reject them because they defy human imagination, and, Platonically speaking, people are governed by plausible imaginations. More agrees and explains that one of the main reasons behind people's resistance to the introduction of new ideas in matter of policy is the claim that abstract and universal arguments cannot be applied to concrete situations.[32]

The crux of the matter is that there is no place for philosophy in politics. More agrees with Hythloday that rulers and politicians turn deaf ears to the *sermo insolens*—the outlandish speech—of the philosophers, for philosophical *sermones* defy common sense and men's willingness to believe. However, More (the character) prefers to rephrase the question by saying that philosophy is useless in politics if it is scholastic, academic philosophy—the philosophy of "universal reason," which "supposes every topic suitable for every occasion." It becomes, therefore, crucial for More (the author) to distinguish between two types of philosophy, between *scholastica philosophia*, which is false and deceiving in its claim that human reason is in the position to reach a universal and necessary knowledge of things, and *civilis philosophia*, which, far from assuming extravagant poses and giving unusual and outlandish speeches, adjusts its arguments to concrete situations (*quae suam scaenam novit*—adapts itself to the drama in hand) and proceeds in a circuitous manner, following a *ductus obliquus*. Hythloday responds to More's antithesis by introducing a third type of philosophy, esoteric and uncompromising, that is, Plato's unwritten philosophical practice.[33] Of the three types of philosophy, though, only the *civilis* one—the public and social one—can be truly persuasive.

Here More the politician seems to be distancing himself from Hythloday the idealist. More is reenacting the Platonic paradox: the domain of politics is refractory to philosophy and urgent action (governed by *prudentia*) cannot be reconciled with theory (enlightened by *sapientia*). The only feasible alternative for the politician is to become a philosopher. Such a solution, though, is so implausible that no one would believe in it. Reason can be persuaded only if the imagination is lured—again in a very Platonic way—to the level of the ideal commonwealth through the use of fictions and myths.[34] The existing society is not real; what is "really real" is the ideal world represented through Utopia. Only cunning stretches of the imagination and fictional suspensions of reality may mitigate the harsh attitude of philosophical disbelief.

Significantly, More the author has More the character in the dialogue introduce the image of the play to make the case of *civilis philosophia* even stronger. Reality is tragicomic and folly is ruling the world. Comedy—indeed comedy of a Plautian kind—is the essence of life. Therefore, if philosophers want to have a say in discussions about reality, they need to be more attuned to the comedy of life. They cannot ignore their audience, and the audience of the comedy of life expects nonsense from the stage (*vernulae nugantes inter se*). Instead to resort to the *insolens sermo* of pure reason, philosophers should adopt a *ductus obliquus*, a more sinuous approach, capable of adapting itself to the varying circumstances of life. The insertion of a tragic scene conveying Stoic moral principles—Pseudo-Seneca's *Octavia*—would not make a comedy more believable or bearable.[35] If we expand on the simile suggested by More, then, we have three political alternatives corresponding to three theatrical genres: *philosophia civilis* is comedy; Machiavellism, that is, the desolate situation of early modern politics, is real tragedy; and Hythloday's idealism, finally, with its Socratic and Platonic subversions of planes—the unbearable injustice of factual reality versus the implausible harmony of imaginary worlds—is sheer tragicomedy.[36]

The Tempest: *Make-Believe as Enchantment*

By using stage metaphors in the passage from *Utopia* mentioned above, More reinforces the view that real life requires circumstantial wisdom and, in a dialectical way, he adds the caveat that one should not pretend to be what he or she is not. From this point of view, the metaphor that life is theater is used precisely to defuse the very threatening aspects of theatricality inherent in human experience. This is certainly not the case with the use of theater within theater that is the defining trait of Prospero's illusions. *The*

Tempest explores the very incantatory nature of theatrical representation.[37] If it is true that a play is somehow a form of collective enchantment, it is also true that an enchantment is a form of theater. It is a point that has been long recognized by theorizers of both magical ritual and theatrical performance. A state of suspended disbelief, close to aesthetic illusion, is required for a spell to become effective. Likewise, the success of theatrical catharsis relies on a condition of vigilant credulousness. A difference between the two, perhaps, is that artistic catharsis—the purging of the affects—is, as Aristotle acknowledged long ago, of a loftier and nobler nature, whereas magical catharsis is meant to satisfy more immediate urges and more prosaic desires, such as cravings for sex, money, health, and physical beauty. In both cases, though, the success of the operation depends on the extent to which the viewer loosens the concentration of the vigilant mind and surrenders to the action represented by the performer or the enchanter.

This aspect of empathetic surrender is lucidly described by Pietro d'Abano, one of the most perceptive authors to have written on the nature of enchantments, who died around 1316 in odor of necromancy. His *Conciliator differentiarum quae inter philosophos et medicos versantur* (ca. 1303) can be considered one of the most powerful and solid synthesis of philosophy, medicine, and magic ever written. In some of its sections, Pietro lays bare the rhetorical and emotional mechanisms involved in magical transactions. He highlights the ability to alter the emotional response of the audience as a key element in the enchanter's procedure. I quote from the *Differentia* 156 (*Utrum praecantatio in cura conferat, necne*—Whether enchantments are an effective cure or not) of the *Conciliator*:

> A speech may assume three forms (see Boethius, *De interpretatione*, I): the first is the process in the mind of the enchanter; the second is a statement according to what Aristotle says in his book on the categories, and this statement can also be uttered in words; the third is a written proposition that makes use of letters and can be hanged around the neck or another part of the body. Moreover, an enchanter needs to be shrewd (*astutus*), self-confident (*credulus*), capable of relying on a wide range of emotions (*affectuosus*), endowed with a powerful and receptive soul (*animae fortis impressivae*); the person to be enchanted must be greedy (*avidus*), full of expectations (*sperans quam maxime*) and in a state of complete receptivity (*dispositus omnimode*). In this way, the action falls on a matter that is predisposed to receive it (see *differentia* 135). And rightly so, because an enchantment, being a process that involves intentional components, is not effective unless the conditions mentioned above are fulfilled, and this is because active principles act on a predisposed subject, which is complying and receptive, and they have the power of altering and transforming what is changeable in the highest degree, that is, the animal faculty, and above all during sleep, because it is during sleep that all the other bodily motions stop.[38]

Let us pause on some of the adjectives used by Pietro: *astutus*, *credulus*, *affectuosus*, and *animae fortis impressivae*. They define the principal features of the enchanter. In addition—and crucially—the enchanter needs a predisposed subject, willing—not necessarily in a fully conscious way—to give up part of his or her imaginations and beliefs. Like the juggler described by Bacon in *Sylva Sylvarum*, who requires an "imaginant" in order to transform the imaginations he has induced in him into beliefs, an enchanter needs a fertile terrain of emotions to prey on. This is precisely the situation that is masterfully depicted in Shakespeare's bewitched island.

The question of the type of magic involved in *The Tempest* is particularly complex and has vexed scholars for quite a long time. The difficulty depends on a number of issues, such as the question of the inherently "magical" nature of all theatrical forms of art; the specific issue of the popularity of plays representing magical subjects in England between the sixteenth and the seventeenth century; and, finally, the question of whether these plays may be considered as representative of contemporary religious, philosophical, and political debates on magic. The view that Prospero practices white, spiritual magic, resonating with Neoplatonic philosophical motifs (theurgy) in contrast to Sychorax's black, demonic magic (goety) is no longer tenable. Prospero's magic is unequivocally "rough," which means it is both unscrupulous and dangerous.[39] In this respect, Prospero acts precisely as the conjuror described by Pietro d'Abano—*astutus*, *credulus*, *affectuosus*, and *animae fortis impressivae*. And like Pietro's enchanter, he is perfectly aware that, to be effective, his charms require an empathetic audience—both on stage and outside the stage. Certainly what Prospero is not is a New Age mage.

Some of Prospero's spells are mere illusions. Some others are real. What is real in all circumstances, though, is the emotional engagement of the participants in the enchanting transaction. In this sense, humanity—no matter whether of the "human, all-too-human" kind—remains the ultimate foundation of the bewitching ritual. Some scholars, in order to emphasize the moral improvement in Prospero's abjuration of magic, have insisted on Plutarchean and Neostoic accents in some of his most noble pronouncements. In fact, what is really at stake here is the realization that Prospero, unlike Ariel, is human and that his spells work because humans' beliefs are inextricably intertwined with affects and imaginations: "Your charm so strongly works'em," Ariel tells Prospero, "That, if you now beheld them, your affections / Would become tender" (5.1.16–17). It is precisely such a tangle of disposition to believe and to imagine that makes his magic effective. A charm cannot work one way only. By recognizing the *afflictions* of his enemies, Prospero's *affections* become tender. He feels he is "one of their kind that relish all as sharply, passion as they." "They shall be themselves," he concludes referring to his prisoners, but he also will be himself at the end of the final act, when readers and viewers are left with an acute sense of disenchantment and disillusion ("PROSPERO: You do yet taste / Some subtleties o'th'isle that will not let you / Believe things certain" [5.1.123–125]).

New Atlantis: *Induced Imaginations*

In *The Tempest* Shakespeare touches upon the effects that a utopian organization of society can have on the government of human appetites. "[N]o kind of traffic / Would I admit," sings Gonzalo in the second Act (2.1.149–150). The scene has been interpreted alternatively either as a parody of communal life to reconfirm the superiority of aristocratic values or as expressing genuine nostalgia for the pristine goodness of the kinship of nature.[40] Both in More and Shakespeare, though, utopian regimens does not necessarily imply an optimistic view of the human nature. More (the character) reminds Hythloday that, precisely because the essence of life is Plautian comedy, "it is impossible to make everything good unless all men are good, and that I don't expect to see for quite a few years yet."[41] Even the Utopians, who believe that true pleasure, both physical and mental, is a naturally virtuous state that derives from following the laws of nature, are aware of the unstable and unsettling condition of human appetites. They reject false pleasures as states of self-deception whereby human beings try to assuage deep feelings of deprivation and fear of want (*timor carendi*). It is only when social order is modeled upon patterns of communal life that fear of penury and consequent drives to acquisitiveness may naturally be converted into social tendencies.

In *The Tempest* delusion and self-delusion, and the ensuing conflicts, are largely due to the outburst of uncontrolled appetites. The precarious resolution reached at the end of the play is largely based on a compromise between detachment (recognition of the true nature of reality behind the illusion) and engagement (suspension of critical resistances and acceptance of the common bond of human nature). In the first part of *Utopia*, Hythloday describes contemporary English society as a situation in which human appetites are out of control. Farmers and workmen are starved to death and the primal appetite of hunger has transformed them into robbers and beggars. Heedless of the situation, the powerful starve the poor because of their appetite of riches, which, in More's interpretation, is, in fact, a symptom of fearful unsatisfied covetousness. An uncontrollable tendency to conspicuous consumption accounts for the dramatic turn in the economic organization of society (not the other way around); consumption, however, is only a palliative—another psychological mechanism of self-deception—to appease the sense of void and the abyss of insatiable greed. As Hythloday warns Cardinal Morton in *Utopia*, "your island... will be ruined by the crass avarice of a few."[42]

In Bacon's reinterpretation of More's Utopia—Bensalem in *New Atlantis*—appetites are finally tamed and sedated. The development that has led to this result is not disclosed, but from what one can argue by reading the story, progress in technology seems to be behind such an extraordinary accomplishment. Unlike Utopia and Prospero's *signoria* in the island, the commonwealth of Bensalem has managed to find long-lasting ways of governing nature's appetites. The government appears to be lasting and reliable, and involves the management of people's appetites as well as the very appetites of matter. Judging from the outward peaceful appearances of the islanders and their ordinary life, we can conclude that harmony does not rely on the *ductus obliquus* of prudential reason, on deception through enchanting words or on brutal coercion by the intervention of supernatural forces. And yet, of the three islands, Bensalem remains the eeriest one.[43]

In *New Atlantis* the stability of the social and political organization rests on the activities sponsored by the "College of the Six Days' Work" or "Salomon's House," which functions as both an academy of arts and sciences and the main governing body. According to the story that is retold in the course of *New Atlantis*, King Solamona, the legendary lawgiver of the nation, established the college nineteen hundred years before. The college is described as "the noblest foundation . . . that ever was upon the earth; and the lanthorn of this kingdom . . . dedicated to the study of the Works and Creatures of God."[44] So far so good: everything is more or less in line with the narrative conventions of the utopian genre. Things become more distinctively Baconian, though, when one pauses to examine his description of Salomon's House's aims:

> The End of our Foundation is the knowledge of Causes, and secret motions of things; and the enlarging of the bounds of Human Empire, to the effecting of all things possible.[45]

What does Bacon mean by secret motions of things? And what is the link between the knowledge of the secret motions of things and the possibility of transforming reality in an infinite number of ways ("the effecting of all things possible")? From other works of Bacon, including the renowned *Novum organum*, we know that Bacon conceived of these hidden motions as a limited number of material appetites (*Novum organum*, II, 48).[46] One of the most general tenets in his philosophy, if not its most general, is the principle that both the natural and the human world are ruled by appetitive drives inherent in matter. As a result, it is through knowing and subduing the appetites of matter that man can have some hope of mastering nature and restoring a type of humankind in control of its appetites—that is, the original paragon of cognitive and moral perfection embodied by the prelapsarian Adam.

Bacon's idea that the motions of the mind and the motions of the bodies mirror each other is based on the fundamental principle that material cupidity pervades all levels of nature and is an ineliminable source of instability for the system. He is of the opinion that both human and natural creativity, both *ingenia* and *conatus,* are arbitrary forces. Just as a "ruler" (*regula*) symbolizes the metonymical remedy against the bewilderment of human knowledge, so custom is the force that disciplines the unruly nature of human action. The new organon (i.e., the reorganization of man's mental faculties) and the new Atlantis (i.e., the new organization of the social order) work in parallel. Accordingly, the chief institution in Bensalem, devoted to the study of the appetites in matter (the "secret motions of things"), is also in charge of the preservation of the political order. Men of science are in power because, by knowing the appetites of matter, they have a better knowledge of the roots of human appetites.

The central role that appetites play in Bacon's philosophy, however, is also the reason that explains why secrecy looms large in Bensalem and that not all sources and types of knowledge can be disclosed. Appetites need to be circumvented in order to be governed, a procedure that More has described as *ductus obliquus* in his *Utopia.* The establishment of the general good can derive only from a shrewd combination of enlightened knowledge and forceful management of private desires. In Bacon's view, appetites are atomic and self-interested; they have no knowledge of the direction of their movements. However, they can be manipulated by all sorts of "invisible hands," of which custom is the most forceful. In this connection, the figure of the "unknown knower"—of the "spy" as it were—is a pivotal character in Bacon's philosophy as a whole, not only in *New Atlantis.* As communication or social bonding could not rely simply on blind appetite, the rulers of the island need to apply a mechanism whereby the multiplicity of self-interested appetites converges toward the realization of a more comprehensive purpose. This plan can be put into effect by resorting to the device of knowing without being known, of seeing without being seen. This use of "invisible" knowledge, of knowing without being known, takes various forms: simple persuasion (people who happen to land on the island are convinced to stay; those who decide to go back home are not taken seriously by their own countrymen once they return to their countries of origin), cunning of reason (people think they are pursuing their interests, but in fact they are unwittingly working for the realization of a bigger plan, as pawns in a bigger picture) and heterogenesis of ends (what appears to be limited in knowledge and selfish in action is, in fact, a stage toward deeper knowledge and more stable order).

The partially secretive nature of the scientific enterprise as is represented in *New Atlantis* mirrors the partially secretive nature of politics. Both the scientific and the political program deal with the constitutively oblique and devious nature of the appetites. Through the interplay of visibility and invisibility, of knowing and not being known, Bensalem appears to be a microcosmic representation of Bacon's universe, in which order is seen as a result of varying degrees of knowledge. From the fellows of Salomon's

House down to the strangers who are storm-tossed in the island we find a hierarchical distribution of knowledge. Salomon's House, the institution at the center of the political, intellectual, and social life of the island, seems to be characterized by a certain degree of invisibility. The select few who belong to Salomon's House are not always visible or recognizable. In Bacon's story they make their appearance only in special circumstances. The best exemplification of the "unknown-knower" situation in *New Atlantis* is the description of the way in which Bensalemites gather information and intelligence from the other corners of the world, for they seem to have struck a balance between secrecy and publicity: they travel to collect knowledge and new inventions, but they work in incognito and avoid being recognized as traders of knowledge. Industrial espionage is the name for such an activity, but Bacon prefers to call his collectors of intelligence with the noble name of "merchants of light."[47]

Conclusions

Early modern theater, magic, and liturgy share the tendency to prey on human imaginations and exploit their disposition to tolerate limited forms of suspended disbelief, a tendency that I have characterized as "controlled credulousness." In order for such a manipulation of the imagination to work, the flow of imaginative energy needs to be left wavering between states of self-estrangement and self-awareness, in a grey area of liminal experience that shuns the extremes of stuporous wonder and rational detachment. Such an orchestration of suspended imaginations requires the presence of what I have called a "master teller." As already mentioned at the beginning of the chapter, although the narrative frameworks of *Utopia*, *The Tempest,* and *New Atlantis* are constitutively polyphonic and multilayered, the action, nevertheless, is seen from the perspective of a principal narrator: Hythloday, Prospero, and the anonymous narrator in *New Atlantis*. In different ways, their authoritative narratives cast light on skewed processes of acquiring and communicating knowledge: as the Socratic art of intellectual midwifery applied through dialectic and irony, as the ability to conjure up knowledge and action through supernatural means and, finally, as the espionage system carried out by the Merchants of Light every time they collect information from other nations without them knowing. Despite their differences, these three forms of securing knowledge are characterized by the fact of being barren of knowledge of their own. They all prey on someone else's knowledge by adopting oblique strategies of persuasion. More recommends a *ductus obliquus*, Prospero resorts to rough magic, the Fathers of Salomon's House operate in situations of invisibility and secrecy. In different ways, their strategies of communication and winning consent rely on the patterns of persuasion and spell-binding described by Pietro d'Abano. Even Socrates' model of dialogical cross-examination hints at moments of mental stupefaction: Plato had famously compared the effect of being exposed to his mentor's arguments to the state of numbness caused by the torpedo fish (*Meno*, 80A).

As I argued at the beginning of this chapter, behind the early modern debate over the legal and philosophical implications of magical practice, there were often concerns about the use of licit and illicit ways of persuading other wills, of exploiting other people's disposition to imagine reality and suspend disbelief.[48] Magic is rejected in *Utopia*, abjured in *The Tempest*, reformed in *New Atlantis*. And yet it remains an integral part of the rhetorical discourse enacted in the three works. The main issue at stake with magic is that it never leads to a reliable form of persuasion, for incantatory rites are based on coercion. However, it is precisely the possibility offered by magic to exploit the protean energy of human imagination that explains Bacon's interest in magical operations. It is certainly not a coincidence that the project of developing a "high kind of natural magic" outlined in *Sylva Sylvarum* ends with the representation of the "magical" organization of Bensalem in *New Atlantis*.[49] For More, persuasion is an operation falling into the domain of rational activities and based on stringent arguments; as such, it cannot result from manipulating other people's imaginations. Utopians share with More the view that, in the end, "madness will yield to reason," for nature is inherently rational.[50] Utopus, the founding father and lawgiver of More's ideal commonwealth, has realized from the very beginning that there is no way of settling religious conflicts when persuasion fails (*si suadendo non persuadeat*); resort to direct violence, threats, superstitions, and fears would be pointless.[51] The citizens of Utopia (as Gonzalo would later paraphrase in *The Tempest*) distrust all artifices of words. They reject seductive practices, all means to distort the natural condition of things, lawyers, cavils, elaborate laws and treaties. They think that "the kinship of nature is as good as a treaty, and that men are united more firmly by good will than by pacts, by their hearts than by their words."[52] This does not mean that Utopians do not have beliefs that hinge upon preternatural or supernatural phenomena. For instance, they believe that the souls of dead people keep visiting their living friends and relatives, "[a]lthough through the dullness of human sight they remain invisible." Their beliefs in the living presence of the dead souls help the Utopians in their everyday life. "Hence they go about their business the more confidently because of their trust in such protectors; and the belief that their forefathers are present keeps them from any secret dishonourable deed."[53] This is the level of suspended disbelief that More concedes to his Utopians.

Precisely to dispel any possible misunderstanding about the level of awareness that must accompany a truly happy life, More adds as requisites for happiness the ability to understand the meaning of life (*conscientia vitae*), a reasonable exercise of bodily pleasures, and the enjoyment of good health. These, say Hythloday in the dialogue, are all necessary conditions in order for pleasure to be something more than mere stupor: "Mere absence of pain, without positive health, they regard as insensibility, not pleasure" (*Nam dolore prorsus vacare, nisi adsit sanitas, stuporem certe non voluptatem vocant*).[54] More's sentence is a clear echo of one of Hippocrates' aphorisms. Significantly, in *De augmentis scientiarum*, Bacon refers to the same aphorism to make sure that his "medicine of the mind" is not mistaken for an exercise in Stoic self-denial.[55] On the contrary, stupor seems to be the condition affecting both Prospero's bewitched castaways and common Bensalemites (Fathers of Salomon's House aside). Jerry Weinberger has noticed that people in New Atlantis look suspiciously too happy, as if they have been lobotomized.[56] The same cannot be said of the stupor induced by Prospero. The various spells that occur in *The Tempest* involve a condition of mutual responsiveness between the enchanter and the enchanted. Change—not necessarily in the form of a radical sea-change—affect all the participants in the incantatory rite of the play, including the viewers. In the end we all feel part of an unsettling situation of restlessness, a situation that, once again, I would call a state of controlled credulousness. Referring to Shakespeare's "peculiarly intense interest in witchcraft," Stephen Greenblatt has pointed out how Shakespeare seems to have "identified the power of theatre itself with the ontological liminality of witchcraft."[57]

Utopia differs from Bensalem and the magical dukedom recreated by Prospero in that More's imagined world is firmly based on the natural condition of wakefulness and alertness that is typical of reason and on the natural kinship that connects reason with the order embedded in the created universe. Utopians resist all forms of mental stupefaction and the imposition of arbitrary conventions. In this sense, Utopia is an imaginary world that in the end reconfirms the reasonable and intelligible nature of the imagination. By contrast, *The Tempest* and *New Atlantis* herald a new age in the history of world-building, in which the imagination is distrusted and its claims to produce plausible universes of meaning and action are severely questioned. In this, *The Tempest* and *New Atlantis* mark the gradual demise of the Platonic synthesis of belief and imagination, which, revolving around norms of ethical reasonableness and ontological intelligibility, had conveyed for centuries a fairly optimistic view regarding the faculty of human imagination and its power to devise meaningful worlds of imaginary reality. It must be said that the characteristic Platonic diffidence toward the imagination, in any of its forms and periodic reincarnations, has in fact never led to a radically skeptical stance about reality, whether ideal or phenomenal. It is rather the imaginary worlds of Shakespeare and Bacon that, by questioning the very existence of intelligible archetypes behind the appearances of things, become collections of disjointed representations with no stable referential ground—a world of transient shadows and dreams, in one case, and a universe of reflections and refractions projected onto nature by minds bewitched with ineradicable idola, in the other.

Notes

1. Stephen Orgel, "Introduction" to *The Tempest* (Oxford: Oxford University Press, 1987), 31–39; Denise Albanese, "The *New Atlantis* and the Uses of Utopia," *English Literary History* 57 (1990): 503–528; David Norbrook, " 'What Cares These Roarers for the Name of King?': Language and Utopia in *The Tempest*," in Gordon MacMullan and Jonathan Hope, eds. *The Politics of Tragicomedy: Shakespeare and After* (London: Routledge, 1992), 20–54; Cynthia Lewis, *Particular Saints: Shakespeare's Four Antonios, Their Contexts, and Their Plays* (Newark and London: University of Delaware Press, 1997), 158ff.; Annabel Patterson, " 'Thought Is Free': *The Tempest*," in R.S White, ed. *The Tempest. William Shakespeare* (Houndmills and London: Palgrave Macmillan, 1999), 123–134; Roland Greene, "Island Logic," in Peter Hulme and William H. Sherman, eds. *"The Tempest" and Its Travels* (London: Reaktion Books, 2000), 138–145; Giuseppe Mazzotta, *Cosmopoiesis: The Renaissance Experiment* (Toronto, Buffalo, and London: University of Toronto Press, 2001) (Chapter 3: "Adventures of Utopia: Campanella, Bacon, and *The Tempest*"), 53–75; Robert Appelbaum, *Literature and Utopian Politics in Seventeenth-Century England* (Cambridge: Cambridge University Press, 2002), 67; Eric Nelson, "Shakespeare and the Best State of a Commonwealth," in David Armitage, Conal Condren, and Andrew Fitzmaurice, eds. *Shakespeare and Early Modern Political Thought* (Cambridge: Cambridge University Press, 2009), 253–270.
2. Stuart Clark, *Vanities of the Eye: Vision in Early Modern European Culture* (Oxford: Oxford University Press, 2007).
3. Greene, "Island Logic." See also Stephen Greenblatt, *Shakespearean Negotiations: The Circulation of Social Energy in Renaissance England* (Oxford: Oxford University Press, 1988) (Chapter 5: "Martial Law in the Land of Cockaigne"), 129–163; Carlo Ginzburg, *No Island Is an Island. Four Glances at English Literature in a World Perspective* (New York: Columbia University Press, 2000), 1–42.
4. Thomas More, *Utopia*, ed. George M. Logan, Robert M. Adams, and Clarence H. Miller (Cambridge: Cambridge: University Press, 1995), 111.
5. Ibid., 107.
6. Ibid., 82–83. See Eric Nelson, *The Greek Tradition in Republican Thought* (Cambridge: Cambridge University Press, 2004) (Chapter 1: "Greek Nonsense in More's Utopia"), 19–48.
7. On the use of logical paradoxes as counterpart to visual deceptions, see Rosalie L. Colie, *"Paradoxia epidemica": The Renaissance Tradition of Paradox* (Princeton: Princeton University Press, 1966); Clark, *Vanities of the Eye*, 107–108.
8. See Brian Stableford, "Science Fiction before the Genre," in Edward James and Farah Mendlesohn, eds. *The Cambridge Companion to Science Fiction* (Cambridge: Cambridge University Press, 2003), 15–31.
9. See Russ McDonald, *Shakespeare's Late Style* (Cambridge: Cambridge University Press, 2006); Raphael Lyne, *Shakespeare's Late Work* (Oxford: Oxford University Press, 2007); Charles Moseley, "The Literary and Dramatic Contexts of the

Last Plays," in M.S. Catherine, ed. *The Cambridge Companion to Shakespeare's Last Plays* (Cambridge: Cambridge University Press, 2009), 47–69.

10. More uses the word *tragicomoedia* in *Utopia* at page 96.
11. See Barbara Mowat, "What's in a Name? Tragicomedy, Romance, or Late Comedy?" in Richard Dutton and Jean Howard, eds. *A Companion to Shakespeare's Works* (Oxford: Blackwell, 2003), 4:129–149. On the material context underlying English early modern interest in tragicomedy, see Valerie Forman, *Tragicomic Redemptions: Global Economics and the Early Modern English Stage* (Philadelphia: University of Pennsylvania Press, 2008).
12. More, *Utopia*, 40.
13. Samuel Taylor Coleridge, "Notes on *The Tempest,*" in Patrick M. Murphy, ed. "*The Tempest*": *Critical Essays* (New York and London: Routledge, 2001), 101–106 (originally published in *The Literary Remains of Samuel Taylor Coleridge*, ed. Henry Nelson Coleridge, 4 vols. (London: 1836–1839), 2:92–102.
14. For a recent discussion of Aristotle's notoriously complex views on the imagination, see Ronald Polansky, *Aristotle's "De anima"* (Cambridge: Cambridge University Press, 2007), 403–433.
15. The classical locus is Avicenna: *"Avicenna," Liber de anima, seu sextus de naturalibus*, 2 vols., ed. Simone van Riet (Louvain and Leiden: Peeters and Brill, 1972), 2:62–66.
16. I hope readers are charitable enough to forgive my cursory remarks on the history of the imagination, but no such history is needed here. My observations are meant only to provide a brief contextualization to the notion of "believable imagination." The bibliography on the subject of premodern and early modern imagination is, of course, imposing. Here are some indications: Elizabeth R. Harvey, *The Inward Wits: Psychological Theory in the Middle Ages and the Renaissance* (London: Warburg, 1975); Frances A. Yates, *The Art of Memory* (London: Pimlico, 1992); Janet Coleman, *Ancient and Medieval Memories: Studies in the Reconstruction of the Past* (Cambridge: Cambridge University Press, 1992); *Phantasia-Imaginatio,* ed. Marta Fattori and Massimo Bianchi (Rome: Edizioni dell'Ateneo, 1998); Mary Carruthers, *The Craft of Thought: Meditation, Rhetoric, and the Making of Images, 400–1200* (Cambridge: Cambridge University Press, 1998); Peter Mack, "Early Modern Ideas of Imagination: The Rhetorical Tradition," in Lodi Nauta and Detlev Pätzold, eds. *Imagination in the Later Middle Ages and Early Modern Times* (Leuven: Peeters, 2004), 59–76; Bernd Roling, "Glaube, Imagination und leibliche Auferstehung: Pietro Pomponazzi zwischen Avicenna, Averroes und jüdische Averroismus," in Andreas Speer und Lydia Wegener, eds. *Wissen über Grenzen. Arabisches Wissen una lateinisches Mittelalter* (Berlin and New York: De Gruyter, 2006), 677–699.
17. Bacon, *Sylva Sylvarum,* in *Works*, 14 vols., ed. James Spedding, Robert Leslie Ellis, and Douglas Denon Heath (London: Longman, 1857–1874),

2:654. See Aristotle, *De anima*, 3:3, 427b 21. On Bacon's views on the imagination, see K.R. Wallace, *Francis Bacon on Communication and Rhetoric* (Chapel Hill: University of North Carolina, 1943); Marta Fattori, "*Phantasia* nella classificazione baconiana delle scienze," in M. Fattori, ed. *Francis Bacon: Terminologia e fortuna nel XVII secolo* (Rome: Edizioni dell'Ateneo, 1984), 117–137; G.T. Olivieri, "Galen and Francis Bacon: Faculties of the Soul and the Classification of Knowledge," in D.R. Kelley and R.H. Popkin, eds. *The Shapes of Knowledge from the Renaissance to the Enlightenment* (Dordrecht: 1991), 61–81.

18. More, *Utopia*, 224–225: "persuasum habeant nulli hoc in manu esse ut quicquid libet sentiat."
19. Ibid., 227.
20. See Heinrich Cornelius Agrippa of Nettesheim, *Opera* (Lyon: Bering Brothers [n. d.]), 1:111. I owe this reference to Allison Kavey.
21. Bacon, *Sylva Sylvarum*, 654. Another hint at Aristotle's *De anima,* 3:3, 427b 18.
22. Bacon, *Sylva Sylvarum*, 654.
23. Ibid.
24. Ibid., 656: "by first telling the imaginer, and after bidding the actor think."
25. Ibid., 654–655.
26. Bacon, *Sylva Sylvarum*, 655:

> I related one time to a man that was curious and vain enough in these things, that I saw a kind of juggler, and that had a pair of cards, and would tell a man what card he thought. This pretended learned man told me it was a mistaking in me; "for," said he, "it was not the knowledge of the man's thought, (for that is proper to God,) but it was the enforcing of a thought upon him, and binding his imagination by a stronger, that he could think no other card." And thereupon he asked me a question or two, which I thought he did but cunningly, knowing before what used to be the feats of the juggler. "Sir," said he, "do you remember whether he told the card the man thought, himself, or bade another to tell it?" I answered (as was true) that he bade another tell it. Whereunto he said, "So I thought: for," said he, "himself could not have put on so strong an imagination; but by telling the other the card (who believed that the juggler was some strange man, and could do strange things) that other man caught a strong imagination." I hearkened unto him, thinking for a vanity he spoke prettily. Then he asked me another question: saith he, "Do you remember, whether he bade the man think the card first, and afterwards told the other man in his ear, what he should think; or else that he did whisper first in the man's ear that should tell the card, telling that such a man should think such a card, and after bade the man think a card?" I told him, as was true, that he did first whisper the man in the ear, that such a man should think such a card. Upon this the learned man did much exult and please himself, saying; "Lo, you may see that my opinion is

right: for if the man had thought first, his thought had been fixed; but the other imagining first, bound his thought." Which though it did somewhat sink with me, yet I made it lighter than I thought, and said, "I thought it was confederacy between the juggler and the two servants:' though indeed I had no reason so to think; for they were both my father's servants, and he had never played in the house before."

27. Francis Bacon, *New Atlantis,* in *The Major Works,* ed. Brian Vickers (Oxford: Oxford University Press, 1996), 484–485. Bacon was, of course, fully aware of the power of theatrical representations. See Brian Vickers, "Bacon's Use of Theatrical Imagery," in William A. Sessions, ed. *Francis Bacon's Legacy of Texts* (New York: AMS Press, 1990), 171–213.
28. Which is nevertheless a very important component of the imagination. See Shaun Nichols: "Philosophy would be unrecognisable if we extracted all the parts that depend on the imagination" ("Introduction" to *The Architecture of the Imagination: New Essays on Pretence, Possibility, and Fiction,* ed. S. Nichols [Oxford: Clarendon Press, 2006], 1).
29. Francis Bacon, *Historia naturalis et experimentalis,* 2:14; 5:132.
30. Francis Bacon, *De augmentis scientiarum,* in *Works,* 1:640–641; 4: 428–429.
31. More, *Utopia,* 82.
32. Ibid., 94–97.
33. Ibid., 94.
34. Ibid., 82.
35. On More's reference to *Octavia,* see David Norbrook, "The *Utopia* and Radical Humanism," in Id., *Poetry and Politics in the English Renaissance,* rev. ed. (Oxford: Oxford University Press, 2002), 23. On the reception of Senecan tragedies in early modern Europe, see Roland Mayer, "*Personata Stoa*: Neostoicism and Senecan Tragedy," *Journal of the Warburg and Courtauld Institutes* 57 (1994): 151–174. For a recent edition of Octavia, see *Octavia. A Play Attributed to Seneca,* ed. R. Ferri (Cambridge: Cambridge University Press, 2003).
36. On the Platonic aspects in More's *Utopia,* see now Jean-Yves Lacroix, *L'"Utopia" de Thomas More et la tradition platonicienne* (Paris: Vrin, 2007).
37. See A.D. Nuttall, *Two Concepts of Allegory: A Study of Shakespeare's "The Tempest" and the Logic of Allegorical Expression* (New Haven and London: Yale University Press 2007), 139, 157–158; Pauline Kiernan, *Shakespeare's Theory of Drama* (Cambridge: Cambridge University Press, 1996) (esp. Ch. 5, 91–126); T.G. Bishop, *Shakespeare and the Theatre of Wonder* (Cambridge: Cambridge University Press 1996), 1–16.
38. Pietro d'Abano, *Conciliator controversiarum, quae inter philosophos et medicos versantur* (Venice: Giunta, 1565) (repr. Padua: Antenore, 1985), fol. 212^{v}b:

"Cum oratio triplex consistat (Boethius, Perhihermeneias, I), una quidem in mente ipsius formatus existens conceptus, alia in pronunciatione iuxta illud praedicamentorum. Dico autem orationem voce prolatam: reliqua vero in scripto, ut quod literis a characteribus etiam promitur, et collo tandem, vel alibi alligatur. Amplius praecantator debet esse astutus, credulus, affectuosus, animae fortis impressivae; incantandus vero avidus, sperans quam maxime ac dispositus omnimode, ut actio concidat in materiam praeparatam (differentia 135). Et merito, quia cum incantatio sit quid tanquam intentionale, non agit efficaciter, nisi interveniant praedicta, cum actus agentium sit in passum et susceptivum praedispositum, et ideo pertransmutat et alterat, quod est maxime permutabile, ceu virtutem animalem, et maxime somniis, cum motus reliqui tunc cessent corporei."

39. On the specific question concerning the kind of magic that is represented in *The Tempest*, see Frank Kermode, "Introduction" to *The Tempest* (London: Methuen, 1954); C.J. Sisson, "The Magic of Prospero," *Shakespeare Survey* 11 (1958): 70–77; Frances A. Yates, *Shakespeare's Last Plays* (London and New York: Routledge, 1975) (Chapter 4: "Magic in the Last Plays: *The Tempest*"), 85–106; ead., *The Occult Philosophy in the Elizabethan Age* (London: Routledge, 2001 (originally pub. 1979) (Chapter 15: "Prospero: The Shakespearean Magus"), 186–192; David Young, "Where the Bee Sucks: A Triangular Study of *Doctor Faustus, The Alchemist*, and *The Tempest*," in C. McGinnis Kay and H.E. Jacobs, eds. *Shakespeare's Romances Reconsidered* (Lincoln and London: University of Nebraska Press, 1978), 149–166; M. Srigley, *Images of Regeneration. A Study of Shakespeare's "The Tempest" and Its cultural Background* (Uppsala: Almqvist & Wiksell International, 1985) (esp. Ch. 7 "Rough Magic," 141–166); Ruth Nevo, "Subtleties of the Isle: *The Tempest*," in R.S. White, *The Tempest. Contemporary Critical Essays* (New York: St. Martin's Press, 1999), 75–96; Virginia Mason Vaughan and Alden T. Vaughan, "Introduction" to *The Tempest* (London: Thomson Learning, 1999), 62–73. On witchcraft, theater and imagination in Shakespeare's works, see Stephen Greenblatt, "Shakespeare Bewitched," in Jeffrey N. Cox and Larry J. Reynolds, *New Historical Literary Studies: Essays on Reproducing Texts, Representing History* (Princeton: Princeton University Press, 1993), 108–135; Susan James, "Shakespeare and the Politics of Superstition," in Armitage, Condren, and Fitzmaurice, *Shakespeare and Early Modern Political Thought*, 80–98.
40. Norbrook, " 'What Cares These Roarers for the Name of King?' "; Heather James, *Shakespeare's Troy: Drama, Politics, and the Translation of Empire* (Cambridge: Cambridge University Press, 1997) (Chapter 6: "How Came that Widow In?: Allusion, Politics, and the Theatre in *The Tempest*"), 189–221.
41. More, *Utopia*, 97.
42. Ibid., 65. On *Utopia* and appetites, see Raymond Wilson Chambers, *Thomas More* (Harmondsworth: Penguin, 1963), 118–137; J.H. Hexter, *More's Utopia: The Biography of an Idea* (Princeton: Princeton University Press), 71–81.
43. For some recent literature on Bacon's *New Atlantis*, see Michèle Le Doeuff, "Introduction" to Bacon, *La Nouvelle Atlantide*, ed. M. Le Doeuff

(Paris: Flammarion, 2000), 7–71; *Francis Bacon's "New Atlantis": New Interdisciplinary Essays,* ed. Brownen Price (Manchester: Manchester University Press, 2002); Anthony Grafton, *Worlds Made by Words: Scholarship and Community in the Modern West* (Cambridge, MA, and London: Harvard University Press, 2009) (Chapter 5: "Where Was Salomon's House? Ecclesiastical History and the Intellectual Origins of Bacon's *New Atlantis*"), 98–113.

44. Bacon, *New Atlantis*, in *The Major Works*, 471.
45. Ibid., 480.
46. Francis Bacon, *Novum organum,* in *The "Instauratio magna" Part II*: *"Novum organum" and Associated Texts,* ed. Graham Rees and Maria Wakely (Oxford: Clarendon Press, 2004), 382–417.
47. On the paradigmatically Baconian situation of "knowing without being known," see J.C. Davis, *Utopia and the Ideal Society. A Study of English Utopian Writing, 1516–1700* (Cambridge: Cambridge University Press, 1981), 108–109.
48. For an excellent up-to-date discussion about the level of reality involved in early modern beliefs about magic and witchcraft, see Edward Bever, *The Realities of Witchcraft and Popular Magic in Early Modern Europe: Culture, Cognition, and Everyday Life* (Houndmills and New York: Palgrave Macmillan, 2008).
49. In *Sylva Sylvarum*, Bacon claims that he is developing a new form of magic: "this writing of our *Sylva Sylvarum* is (to speak properly) not natural history, but a high kind of natural magic" (*Works*, 2:378). And he continues: "For it is not a description only of nature, but a breaking of nature into great and strange works."
50. More, *Utopia*, 225.
51. Ibid., 222.
52. Ibid., 201.
53. Ibid., 227. On the religion of the Utopians, see Richard Marius, *Thomas More: A Biography* (Cambridge, MA: Harvard University Press, 1984), 171–183.
54. More, *Utopia*, 174.
55. Bacon, *De augmentis scientiarum*, in *Works*, 1:732: "quin potius cum illo Hippocratis aphorismo concludimus: *Qui gravi morbo correpti dolores non sentiunt, iis mens aegrotat.* Medicina illis hominibus opus est, non solum ad curandum morbum, sed ad sensum expergefaciendum." See Hippocrates, *Aphorisms*, 2:6.
56. Jerry Weinberger, "On the Miracles in Bacon's *New Atlantis,*" in Bronwen Prices, ed. *Francis Bacon's New Atlantis: New Interdisciplinary Essays* (Manchester and New York: Manchester University Press, 2002), 106–128 (107). See also id., "Introduction" to Francis Bacon, *New Atlantis and The Great Instauration,* ed. J. Weinberger (Wheeling, IL: Harlan Davidson, 1989), vii–xxxiii; id., "Francis Bacon and the Unity of Knowledge: Reason and Revelation," in Julie Robin Solomon and Catherine Gimelli Martin, eds. *Francis Bacon and the Refiguring of Early Modern Thought: Essays to*

Commemorate "The Advancement of Learning" (1605–2005). (Ashgate: Aldershot, 2005), 109–127.

57. Stephen Greenblatt, "Shakespeare Bewitched," in Jeffrey N. Cox and Larry J. Reynolds, eds. *New Historical Literary Studies. Essays on Reproducing Texts, Representing History* (Princeton: Princeton University Press, 1993), 120.

CHAPTER 6

Imagination and Pleasure in the Cosmography of Thomas Burnet's *Sacred Theory of the Earth*

Al Coppola

The work of the imagination is central to the cosmological enthusiasm of the late seventeenth century. In response to new discoveries of natural philosophy, from the new astronomy of Copernicus and Galileo that decisively unseated humankind from the center of the universe, to new researches into the natural history of the earth, there was an urgent need to produce new cosmographies that were both scientifically coherent and theologically palatable. This chapter considers the most influential popular cosmology of late seventeenth-century England in order to chart the trajectory along which cosmology as a serious, physicotheological pursuit increasingly shifted to a form of popular science education and mass entertainment.

Thomas Burnet's *Sacred Theory of the Earth* (1681/1689) appeared just a few years before the other great work of seventeenth-century popular cosmology, Bernard Fontenelle's unrepentantly delightful *Entretiens sur la Pluralité des Mondes* (1686). On the surface, these two books appear to be quite different: the former attempts to scientifically verify the Bible's account of Creation and prove that all the changes that earth has experienced and will undergo, from the moment of its creation to its prophesied conflagration, can be explained by fixed universal mechanical laws that govern the properties of all matter and motion. The latter, notable for its almost complete neglect of religion, attempts to teach the new astronomy and Cartesian physics in the form of a series of learned and witty conversations between a philosopher and a beautiful Marquise, who sit up in the noblewoman's garden at night to gaze at the stars and marvel over the natural philosophic knowledge to be gained from an enlightened appreciation of the spectacles they behold. Fontenelle's work was almost universally hailed, both beloved by its readers and approved by the experts who steered the Enlightenment's science popularization initiative. Its power lay in its extraordinary appeal to the readers' imagination, conjuring a cosmography so rich, so detailed, and so seductive that, far from disrupting the work's scientific value, it ensured that its scientific knowledge would be seamlessly accepted. Burnet's cosmography also exploited the power of the imagination to figure forth its scientific and theological truths, and readers were drawn to what seemed like Burnet's seamless synthesis of new physics and Mosaic history. But after the initial wonder and applause died down, it was precisely the imaginative audacity of Burnet's work that left it vulnerable to critique.

Indeed, the fate of Burnett's work, first published just six years before Fontenelle's, was vastly different. The imaginative qualities that made Burnet's cosmography seem so persuasive and so religiously scrupulous to its first readers were the very aspects of the work that led critics to attack both its scientific and theological rigor, and the text entered into a long period of critical disfavor. However, even after it was conclusively discredited both as a work of natural philosophy and as a work of theology, Burnet's text enjoyed a second life in eighteenth-century belles-lettres, as a "philosophic romance" par excellence and a model for the kinds of "pleasures of the imagination" that Joseph Addison was to later champion in the *Spectator.* Although little-studied now, Burnet's *Sacred Theory* was a critical touchstone for the late seventeenth and early eighteenth centuries in England, and this chapter seeks to analyze the work's extraordinary trajectory from serious physicotheology to edifying philosophical romance.

The modern critical literature that has taken notice of Burnet's *Sacred Theory* has generally viewed it as a key event in the renegotiation of the relationship of scientific and religious authority in the seventeenth century. When new knowledges unseated long-held orthodoxies about the nature of the universe, the age of the earth, and the physical laws governing matter—new truths that were not inevitably conformable to the accounts of such matters delivered in the Bible—there emerged an urgent need to remake the old cosmologies. It has been a critical commonplace in the history of ideas since at least mid-century that "world-making was, in the 1690's, the central issue in the struggle of science and religion."[1] While more recent work has tended to complicate the simple binary implied in the notion of a "struggle" between religious and natural philosophic truth-claims, Burnet's *Sacred Theory* has remained central to recent work assessing the role of Protestant theology in the development of early modern natural philosophy.[2] Attention has specifically focused on the interpretive liberties that Burnet took with the biblical accounts of the Creation, the Flood, and the the Last Judgment.

As part of his provocative thesis that a new literalism in Protestant biblical exegesis conditioned a later shift of focus on the "things themselves" in natural philosophy, Peter Harrison has identified the Burnet controversy as a crucial friction point in the late seventeenth-century redefinition of the longstanding Christian exegetical tradition of accommodation. Long before Burnet penned his cosmography, Christian theologians had affirmed that a literal reading of the scripture did not preclude the allowance that some aspects of the Bible were purposely expressed in terms that were familiar to mankind's limited human intelligence or to the vulgar notions of its first readers.[3] According to Harrison, however, what *was* new in theorists like Burnet was an expectation that the Bible contained latent scientific, philosophic, and historical truths—specifically, in Burnet's case, that "the Bible contained the suppressed mechanics of Creation underneath its textual surface."[4] For Burnet and other like-minded physicotheologians, the full range of knowledge contained in the Bible became a help to, as well as a check upon, the production of true natural philosophy. This meant not just that Mosaic history was being given a fresh look for its possible scientific relevance; rather, "discoveries in the book of Nature were [believed] to shed light on the neglected scientific treasures of the sacred text."[5] That is, the new science was thought to provide a means for discerning heretofore unrecognized biblical wisdom in matters natural philosophical. While none of the actors in this controversy ever proclaimed to be doing anything except employing a sober literalism in their analysis of either the things of nature or the things of scripture, it is clear that the imagination, in certain circumstances, came to have a radically expanded interpretive role in both theology and natural philosophy in the late seventeenth century.

Recent critical work has located Burnet in a larger narrative of changes in the nature of exegetical literalism and accommodation, but the analysis here asks a smaller, more stubborn question: why should this controversy have played out around the specific question of cosmography? It was in the "making of worlds," in Tuveson's phrase, that these physicotheological controversies coalesced, but the implications of this fact have not been fully explored. Nor has the question of why Burnet's *Sacred Theory,* which was definitively discredited on both theological and scientific grounds by the end of the 1690s, should have enjoyed a remarkable popularity across the eighteenth century, on par with that of the other great work of imaginative cosmology and (scientific) accommodation to emerge in the 1680s, Fontenelle's *Entretiens.*

The best way to delve deep into these problems is by attending more closely to the aspect of Burnet's work that critics consistently harped upon: the allegedly fictional quality of his world-making. In the eyes of his contemporaries—both his critics and his admirers—his text offered itself up as a "romance" for imaginative consumption, in spite of (although Burnet would say, because of) the text's purported philosophical and theological rigor. In light of this, we might helpfully take a cue from Stephen Jay

Gould, who has explicated the *Sacred Theory* as primarily a triumph of narrative, wherein Burnet knowingly exploits "the natural proclivities of human curiosity" to enlist the reader in his theory.[6] For Burnet, world-making *is* fiction-making, and paradoxically this is ultimately the most effective form of both scientific and religious truth-telling.

"A World Lying in Its Rubbish": Contexts of Loss for Burnet's Cosmography

Thomas Burnet published the first part of *Telluris Theoria Sacra* in Latin in 1681. One of the most widely read and debated works of late seventeenth-century physicotheology, the text was initially greeted with great interest and applause by a wide segment of polite culture in the Restoration that was as enthusiastic about the new science as it was cautious to reconcile the newly discovered truths about the natural world with orthodox Christianity. Within a few years of its publication, and particularly after it was translated in 1684 out of Latin and into the King's English at the explicit command of Charles II, Burnet's theory was caught up in a gathering storm of controversy. It drew fire from churchmen concerned with the liberties Burnet took in the interpretation of Mosaic history, as well as from natural philosophers who doubted the scientific rigor of the bishop's theory. Burnet's *Sacred Theory of the Earth* placed the problem of cosmography in a post-Copernican and post-Cartesian world at the center not just of learned debates about physicotheology, but also of polite conversation as well, precipitating a deluge of responses—if the pun can be excused—from commentators and animadverters, as well as prompting a series of new theories from writers like William Whiston and John Woodward who sought to correct Burnet's apparent mistakes but nevertheless furthered his cosmological project.

Burnet's theory, in essence, was founded upon the Christian belief in a triple world. Drawing on a range of Old and New Testament sources, this tradition asserted that the earth was first created perfect by God as described by Moses in Genesis 1:3; that after the flood of Noah the very face of the earth was remade anew; and that as prophesied by the New Testament authors, the world will be remade yet again into the "new heaven and new earth" that will survive for a millennium before the earth's final annihilation in the Last Judgment. In drawing on this belief to frame his cosmography, Burnet made two assumptions: (1) that God made the earth perfect at first; and (2) that the processes by which the world was created—and by which it has and will undergo its subsequent transformations—must have proceeded according to the (recently rediscovered) immutable physical laws of nature.

Accordingly, Burnet posited that at first the world emerged out of a watery chaos by the settling of its constituent parts according to their specific gravity: that is, the heavier waters formed a subterranean abyss, and a perfectly flat, smooth, and uniform crust settled on top of it, much like how fine particulates will form a scum upon an oily surface. Perfectly positioned on its axis perpendicular to the sun, this perfectly spherical egg-like earth enjoyed an eternal spring in its first days, as the sun caused what waters were necessary for life to be evaporated out of fissures in the crust. In this scheme, the biblical flood transpired entirely due to mechanical causes, as the interior waters slowly expanded from the heat, eventually rupturing the crust and causing the abyss to break forth and drown the world. The world that we know, then, is merely the broken ruins of the uniform and perfect original earth. Such a synopsis does only the faintest justice to Burnet's ingenious theory, which renders this highly speculative natural history of the earth in far more detail and complexity, and which (in the second volume published eight years later) goes on to demonstrate how these same physical processes will lead inexorably to the prophesied millennium and annihilation. Considering the outlandishness of the theory as presented here in brief, the sophistication of Burnet's explication was considerable indeed to make it seem so compelling for his first readers.

Notwithstanding the serene majesty of the *Sacred Theory,* we must not overlook the fact that Burnet's text made its appearance in a singularly unsettled moment in an already tenuous and tumultuous period of English history. The first flush of enthusiasm for the Restoration of the monarchy had long since passed, as ideological, economic, and political divisions—many unresolved from the Civil Wars and Interregnum, along with a good many new ones—had broken out in full fury with the Popish Plot of 1679. By 1681, the English church and state were embroiled in the Exclusion Crisis, which was then at its height. It was in this cultural milieu that Burnet's *Sacred Theory* first appeared, and first won the patronage of the beleaguered king. I'd suggest that it is with this in mind that we must reckon with Burnet's curious rhetoric of ruin and loss in a text ostensibly devoted to tracing the sure hand of the divine architect, as well as his text's investment in the imaginative recreation of perfect lost worlds.

It is telling to consider Burnet's 1684 dedication to King Charles, which begins the first English translation. In it, Burnet takes pains to figure his text as itself a world, metonymically equating his book with the earth that it reimagines and whose whole natural history is explained in his cosmology. Burnet gifts this "world" to his royal patron, but he explicitly figures this gesture as an act of recuperation, a (re)calling forth of a perfect world that was now "past and gone," a gesture that is offered as an act of recompense for the "unshapen earth" that both the king and the churchman now inhabit. "New-found Lands and Countries accrew to the Prince, whose Subject makes the first Discovery," begins Burnet, but his dedication brings fresh poignancy to this stale rhetorical convention of the form.[7] He continues:

> Having retrieved a world that had been lost, for some thousands of Years, out of the Memory of Man, and the Records of Time, I thought it my Duty to lay it at your Majesties Feet. 'Twill not enlarge your Dominions, 'tis past and gone; nor dare I say will it enlarge Your Thoughts; But I hope it may gratifie your Princely curiosity to read the Description of it, and see the Fate that attended it.[8]

The "world" that the ingenious Burnet has recovered is, admittedly, no South Sea trading post or any other thing of material value that might benefit his king politically or economically in these straightened times. There is a palpable sense of mournfulness here over the enormity of what has been lost, as well as a rueful conviction that the very procedure of history-writing is wholly insufficient—with all memory of the antediluvian world "gone" from the "Memory of Man," the reader is left to ponder how more mundane and recent cataclysms will fare in the "Record of Time." Yet Burnet imagines that his remaking of the world would nevertheless serve to "gratify" the long-suffering prince, who by entering into Burnet's cosmological project would draw pleasure both from the vivid "Description" as from the moral to be drawn from the "Fate" of that first, perfect world.

All earthy things are subject to decay, Burnet seems to be implying, from the frame of the globe to the succession of monarchs. All that remains, according to Burnet, are the fragments that signify the absence of the antediluvian whole:

> We have still the broken Materials of that first World, and walk upon its Ruines; while it stood, there was the Seat of Paradise, and the Scenes of the Golden Age: when it fell, it made the Deluge; And this unshapen Earth we now inhabit, is the Form it was found in when the Waters had retir'd, and the dry Land appeared...

The persistence of the rhetoric of ruin and loss in Burnet's cosmography is effectively required by the specific character of his physicotheology, insofar as it presupposes the world to have been created perfect in form as well as perfectly in accordance with self-perpetuating mechanical laws, a world of pristine simplicity that was only subsequently confounded into the irregular, often inhospitable and frequently hostile world of historical time. Yet this dedication, alongside other select moments in his text, suggests that this rhetoric of ruin and loss is part of Burnet's strategy to propose the imaginative work of his cosmography as a distraction from and/or displacement of certain troubling events and memories of his own historical time—a higher contemplation of the world that takes one's mind off the troubles of this world. Thus Burnet asks, secure in the answer, "what subject can be more worthy the thoughts of any serious person, that to view and consider the Rise and Fall, and all the Revolutions, not of a Monarchy or an Empire, of the Grecian or Roman State, but of an intire World?" (5). Cosmography becomes an escape into the imagination, a refuge from uncomfortable reflections upon actual monarchs and empires rising and falling.

In keeping with this strategy, Burnet's cosmography tends to erase the importance and assert the sameness of all the natural and human history that has passed between the great geologic epochs he posits, from the Creation to the Flood to the Last Judgment. Indeed, he writes "since the Deluge, all things have continued in the same state, without any remarkable change" (326). Because he radically foreshortens the role of both human and divine activity in his cosmography, Burnet places the emphasis on the omnisciently framed (but independently active) mechanical laws of his world, whose history is still unfolding according to the predetermined plan. His cosmography maps not only the forgotten prehistory of the world (which has fallen out of the "Memory of Man"), but also the whole inexorable course of natural events that are yet to befall the world. In what comes to resemble a supreme act of compensation, Burnet pitches his *Sacred Theory* as nothing less than offering a whole "course of nature, or system of natural providence" capable of divining the causes by which the world rose and changed, but also of explaining "what changes it would successively undergo, by the continued action of the same Causes that first produced it" (2). It is as if Burnet is suggesting that knowing the world's system of natural providence/natural philosophy would serve as an imaginative cure for melancholy, since to know the true cosmography is to know that all the world's systems unwind themselves beyond the control of any human effort.

"A Plot Pursued through the Whole Work": Imagination and the Framing of Worlds

Born in ruin and melancholy, Burnet's commitment to what he called a "system of natural providence" is, in fact, the distinguishing feature of his work for critics and enthusiasts alike. His eighteenth-century readership may not have approved of the theological or philosophical particulars of his theory, but it responded most favorably to Burnet's leap of faith in a systematic order to the natural history of the world and to the human history that has transpired within it. What Burnet's cosmography demanded, and what readers seem to have found especially attractive to enter into, was a willingness to "think big" and intuit grander arcs of order and connectedness in their world. In short, to realize its cosmography and make its religious and scientific truths legible, the *Sacred Theory* asked its readers to exercise a well-developed faculty of the imagination capable of enlacing with Burnet the lines of apparent confusion into unified order.

Stressing the need for a specific set of cognitive skills and aptitudes, Burnet insists that his readers must not have a "mean and narrow Spirit" (a2r) where "Souls that are made little and incapacitous cannot enlarge their thoughts to take in any great compass of Times or Things" (a2v). By experience he's found that some auditors simply lack the requisite mental capacities to understand his theory, but what truly vexes Burnet are those who refuse to credit it, "not so much... from the narrowness of their Spirit and Understanding, as because they will not take time to extend them" (a2v):

> I mean Men of Wit and Parts, but of short Thoughts, and little Meditation, and that are apt to distrust every thing for a Fancy or Fiction that is not the dictate of Sense, or made out immediately to their Senses. Men of this Humour and Character call such Theories as these, Philosophic Romances, and think themselves witty in the expression. They allow them to be pretty amusements of the Mind, but without Truth or reality... (a2v)

The "Wits" who criticize him would have him confine his theory to mere "sense"—empirical and immediately verifiable knowledge—yet what is wanting in a good theory, as well as in the "spirit and understanding" of these wits—is, in fact, an imaginative dilation of the mind's critical faculties. These scoffers would dismiss Burnet's work as nothing but a "Philosophic Romance," but what is interesting here is that Burnet is not entirely uncomfortable with such a notion. Indeed, he launches into a defense of his treatise on the grounds that its sweeping synthesis of science and theology is not only securely grounded, but that it is, in fact, all the more truthful for having those very qualities of romance that the foolish, narrow wit would deride or patronize:

> ... Where there is variety of Parts in a due Contexture, with something of surprising aptness in the harmony and correspondency of them, this they [the Wits] call a Romance; but such Romances must all Theories of Nature, and of Providence be... There is in them, as I may say so, a *Plot or Mystery* pursued through the whole Work, and certain Grand Issues or Events upon which the rest depend, or to which they are subordinate; but these things we do not make or contrive ourselves, but find and discover them being made already by the Great Author and Governour of the Universe. (a2v–a3r).

In short, what Burnet has produced in his theory is analogous to a well-plotted play. The *Sacred Theory* is no mere "contrivance" but rather a work of imaginative "discovery," which like a grand and "satisfying" drama, turns upon a wide "variety of Parts" moving together with "surprising aptness" and "harmony." These various parts—exhibiting what playwrights might have called sufficient business and affecting characterization—depend upon "certain Grand Issues or Events," that is, something very much like the unifying action so prized in neoclassical drama. Most importantly, according to Burnet, a great work of science must have "a Plot or Mystery pursued through the whole work" whose very elegance of composition betrays the hand of God—"the Great Author" in the drama of Burnet's Christian cosmography—just as surely as a well-plotted play betrays the steady, guiding artistry of the master playwright.

In these passages, Burnet sounds remarkably like John Dryden in the *Essay of Dramatick Poesie* (1668), when he champions the apparently irregular form of tragicomedy. Dryden vindicates heroic dramas like his *Conquest of Grenada* on the strength of their distinctive "variety," which "if well ordered, will afford a greater pleasure to the audience" than a play that strictly observed the classical rules of unity and decorum.[9] Indeed, the judicious framing of a heroic drama, whose main action is offset by a profusion of subplots, derives its greatest power to please, he says, from its very copiousness, which must still contain, for all its apparent confusion, a clear "Plot or Mystery pursued through the whole work." Describing an experience not so far removed from that of reading Burnet's *Sacred Theory,* Dryden says:

> If then the parts are managed so regularly that the beauty and variety of the whole be kept entire, and that the variety become not a perplexed and confused mass of accidents, you will find it infinitely pleasing to be led in a labyrinth of design, where you see some of your way before you, yet discern not the end till you arrive at it.[10]

Following through with this logic, Burnet is at pains, at almost every turn, to render his theory, not just plausible, but palpably true, by the deployment of spectacular descriptive imagery that is "clearly discover'd, well digested, and well reason'd in every part," handled with all the elan of a master playwright (a3r). In the *Sacred Theory,* Burnet's announced goal was "to *view* in my mind, so far as I was able, the Beginning and Progress of a Rising World" (1, my emphasis).[11] Noting that "'Tis now more than Five Thousand years since our World was made," Burnet gushes that "it would be a great pleasure to the mind to recollect and view it at this distance those first Scenes of Nature, what the face of the Earth was when fresh and new, and how things differ'd from the state we now find them in," betraying a hunger to palpably apprehend by sight those first "Scenes." However, "the speculation is so remote, that it seems to be hopeless," he writes, lamenting, "by what footsteps, or by what guide can we trace back our way to those first ages and the first order of things?" (7).

Trace the way back his text did, however. To read this volume is to be constantly exhorted to "see," to "view," to "behold" the cosmological truth Burnet set about proving. Indeed, the book's most remarkable set piece—his description of the deluge—can helpfully serve as an exemplum for a rhetorical strategy that Burnet pursues throughout the book. In this famous, and famously spectacular, "conjectural" description of the flood, enlivened with all the affecting detail he can muster, Burnet envisions how the flood would have been experienced by an immediate observer. He introduces it in the eighth chapter of his first book: although he claims that he has already indisputably "shown" his theory to be a "real account" (96), he nevertheless says that he "would willingly proceed one step further" and "descend to particulars" and give an exact and copious, minute by minute, blow by blow, visualization of the event. While Burnet issues the caveat that, "I desire to propose my thoughts concerning these things only as conjectures, which I will ground as near as I can upon Scripture and Reason," he carries this plan forward, despite the risk of appearing excessively credulous, or improperly hortatory, because just such a "conjectural," "particular" account would best "expose the Theory to a full trial." Even though he has already deductively proven his theory, Burnet insists that it is precisely this kind of fictional experiment—that is, imaginative world-making—that is the best way to "secure assent" to his physicotheology (96).

The particulars that Burnet descends to here are particular indeed—breathless, really—in their intimacy and immediacy, with Burnet pausing at every turn of the story to exhort his readers to see, to feel, to imagine—in short, to put themselves in the affective state of an eyewitness wholly consumed with the unfolding spectacle of the deluge. Just a small sample of the description will serve to give some sense of the power of Burnet's prose here:

> Thus the Flood came to its height; and 'tis not easie to represent to ourselves this strange Scene of things, when the Deluge was in its fury and extremity; when the Earth was broken and swollow'd up in the Abysse, whose raging waters rise higher than the Mountains, and filled the Air with broken waves, with an universal mist, and with thick darkness, so as Nature seemed to be in a second Chaos; and upon this Chaos rid the distrest Ark, that the bore the small remains of Mankind. (99)

The concrete details press upon the reader, marshaled with such a copia of descriptive amplifications ("raging waters," "broken waves," "universal mist," "thick darkness") that this literally unseeable event is nevertheless represented in vivid, immediate detail. Underscoring the importance of the visual in his spectacular cosmography, Burnet concludes his narrative of the deluge with a large engraving that depicts the face of the earth as it looked from a birds-eye, or rather a God's-eye, view (figure 4.1). This illustrative plate—which is actually one of a series commissioned for the work that supply a pictorial representation of each phase of the world in Burnet's cosmography—appears almost comically literal to a modern sensibility in its attempt to unambiguously represent the truth of Burnet's theory and to render those truths self-evidently present to his readers. Indeed, the pictures place the reader in the position of the only possible spectator of these true events, God himself, thereby simulating an omniscient perspective in the mind of Burnet's reader. In the case of the deluge illustration, Burnet renders all the relevant "particulars" of the deluge—"the broken and drown'd Earth," "the floating Ark," "the guardian Angels"—in order to ensure that his theory appears not just true, but pleasurable. He is being quite literal when he says to the reader, about this plate, that "We may *entertain* ourselves with the consideration of the face of the Deluge" (100, my emphasis). Indeed, Burnet would seem to be as much invested in proving the truth of a particular physicotheological theory, as he is in inculcating a particular mode of visually apprehending the world in order to discover its implicit system, which ultimately is the real essence of his cosmography.

Moses His Blockheads: Critical Attacks Scriptural

The imaginative quality of Burnet's *Sacred Theory* began attracting serious criticism by the end of the decade, which was quite a few years after the publication of the book. The first round of attacks was religious in nature, with the most influential and sustained criticism coming from Erasmus Warren, who issued *Geologia: Or Discourse Concerning the Earth Before the Deluge* (1690), and subsequently drew Burnet into a pamphlet war that sustained itself through a handful of separately printed comments, defenses, and animadversions.[12] Warren was far from a reflexively hostile critic: he assumed that a physicotheological cosmology was not only possible but necessary; he only disagreed with the license he felt Burnet was taking in the interpretation of scripture. "Considering it simply and abstractly in itself, as a Philosophic *Scheme* or representation of things," Warren wrote of Burnet's *Sacred Theory,* "I found it a Treatise, not unworthy of the ingenious Author of it"(A2r). "It abounds with Philosophy indeed" continues Warren, "but it is not justly regulated, and kept within due Limits. For it runs so fast, and is driven so far, that it treads unseemly, and insufferably too, upon the heels of... that most Divine and Infallible Truth which was spoken by GOD" (42).

In response to Warren, Burnet flatly denies that his text abuses scriptural authority. Indeed, at times he puts himself across as a strict and unimaginative literalist, noting, for example, that it is impossible to doubt the coming of a new heaven and a new earth since this is upheld by the "plain and genuine sense of the apostle's discourse."[13] However, when it comes to the considerably more thorny question of the truth value of the natural philosophic knowledge conveyed in Moses' account of the Creation in Genesis, Burnet insists that it would be a blatant mistake to take the biblical account literally, as the hexameron was expressly written in a manner to make it more persuasive to its first audience: "The thing itself bears evident marks of an accommodation and condescension to the vulgar notions concerning the form of the world... it is easily reconcilable to both [Nature and Divine Revelation], if we suppose it writ in a Vulgar style, and to the conceptions of the People" (44–45).

As Peter Harrison has discussed, the notion of exactly what might count as a "literal" interpretation of the Bible was a fundamental and deeply felt matter of dispute, and the late seventeenth-century's transformation of notions of literalism emerge, in his book, as one of the watershed epistemological crises of an emergent modernity.[14] Burnet, as he explains, was not necessarily beyond the pale in his claim to be a literalist exegete while nevertheless assuming that the hexameron—at least when it touches on matters of natural history—is something less, or more, than a strictly literal account. However, whether he was provoked by criticism of his *Sacred History* or guided by a profound heterodoxy that he later felt at liberty to disclose only after his chances for preferment in the Church were given up for lost, Burnet was to voice a far more strident criticism of the veracity of Mosaic history in his *Archeologia Philosophicae* (1692), which frankly ridiculed the notion that Mosaic history should be understood as anything but a fable:

> That the account given by Moses, not only of the Origine and Creation of the World, but of Adam, and the first Transgression, and the serpent, and the cursing of the Earth, and other matters relating to the Fall, is not true in it self, but only spoken popularly, to comply with the dull Israelites, lately slavish Brickmakers, and smelling strongly of the Garlick and Onions of Egypt. To humour these ignorant Blockheads that were newly broke loose from the Egyptian Taskmasters, and had no sense nor Reason in their thick sculls, Moses talks after this rate; but not a Syllable of Truth is in all that he saith.[15]

These are Burnet's words here, but this shocking quotation is taken not from Burnet's Latin treatise, but from the churchman John Edwards' outraged indictment of Burnet's views in his *Discourse Concerning the Authority, Stile and Perfection of the Books of the Old and New Testament* (1693), which translated and reproduced the foregoing quotation as a kind of Q.E.D. Edwards' attack on Burnet comes at the beginning of a larger survey of recent abuses of biblical interpretation, in which Edwards asserts that the Bible, and specifically "the beginning of the Book of Genesis... must not be turned into mere Mystery and Allegory: for it is sufficiently evident that Moses speaks of Matter of Fact" (34). Burnet, whose *Archeology of*

Philosophy provides Edwards with his most recent, most high-profile, and most scandalous example, "cannot be rebuked enough got his attempt of turning all the Mosaic History concerning *Adam* and *Eve*... into Parable, yea Ridicule" (34–35).

Edwards' chief complaint, like that of Warren, is that Burnet's cosmography is insufficiently literal; essentially, he accuses Burnet of simply making things up. Burnet may claim that Moses was a fabulist, but it is Burnet himself, according to Edwards, who is the chief spinner of fables, one who indulges a most untoward imaginative license. True, Edwards concedes that "the Scripture sometimes speaks... after the manner of men, and in compliance with their common (though mistaken) Apprehensions." But in the *Archeologia Philosophicae,* and especially when he insults the Israelites as "ignorant Blockheads" "smelling strongly of the Garlick and Onions of Egypt," Burnet "here stretches this too far," he says. Outraged by the gratuitous insults hurled at God's chosen people, but more by the implications that this attitude would have for securing the truth of the Bible's accounts generally, Edwards says that "This is very strange Language from a Reverend Divine, who thereby destroys the whole System of Theology, and Christianity itself" (35–36). In this "Allegorical Gentleman's" imaginative world-making, "all is mere Romance and Fiction" (35). If we were to buy into Burnet's cosmology, Edwards' argument goes, then the gospels "are but a cunningly devised Parable; they may have some *moral* meaning, as *Esop's Fables* have, but they contain nothing of real Fact. This is the natural result of *allegorizing* the 3d Chapter of *Genesis*" (37).[16]

As Edwards' argument makes clear, it is the imaginative—or, as he says, "allegorical"—quality of Burnet's project that is most dangerous. Not only has he "overturn'd, the Foundations of Religion," Edwards complains, the sweetness of his rhetorical performance "hath abundantly gratified the whole Tribe of Atheists and Deists... and indeed, we cannot but observe what sort of Men they are that applaud his Undertaking, *viz.* the Wits of the Town (as they are call'd)... This Theorist has become much more pleasing to them than Mr. Hobbs" (36–37). Indeed, some of the most controversial chapters of *Archeologia Philosophicae* were opportunistically translated and republished by the Deist Charles Blount, who recognized in Burnet a fellow-traveler who discredited the whole notion of a deity actively involved in natural (and by extension, human) history.[17] In the context of these developments, Edwards' own choice of rhetoric is striking: Burnet "pleases," "gratifies," and garners applause, suggesting that what Burnet has really done with his philosophical romance is that he has raised and satisfied an appetite for a well-plotted story, rich in luxurious detail, that inexorably follows a narrative arc of matter in motion with no need for any divine intervention. To the outraged churchman Edwards, this does not show any theological or philosophical rigor on the part of the author, only "his little Talent of Jesting and Drolling" (38–39).

More the Orator Than the Philosopher: Critical Attacks Philosophic

Criticism against Burnet's *Sacred Theory* on theological grounds may have peaked in 1693, but the rest of the decade saw increasing dissatisfaction with the quality of Burnet's science.[18] Other natural philosophers, sympathetic to the aspirations of Burnet's cosmography but disbelieving his explanations, had proposed alternative theories of the creation and deluge, William Whiston's *New Theory of the Earth* (1696) and John Woodward's *Essay toward a Natural History of the Earth* (1695) being the most influential. Both Woodard, the professor of physic at Gresham College, and Whiston, then a fellow at Cambridge's Clare College, identified serious scientific flaws in the *Sacred Theory,* but only as part of an effort to replace Burnet's flawed cosmography with a more plausible exercise in world-making. For all the renewed cosmological enthusiasm in scientific circles, however, it is instructive to read John Keill's definitive demolition of Burnet's scientific credibility in 1698, in a withering Newtonian critique of Burnet's clumsy Cartesian physics. In Keill's response to the deluge controversy, *An examination of Dr. Burnet's Theory of the Earth, together with some remarks on Mr. Whiston's New Theory of the Earth,* this prominent natural philosopher—no backwater reactionary, having offered the very first lectures in Newtonian physics at Oxford—attacks Burnet (and, somewhat more gingerly, Whiston) for writing what he too calls "romance," and in doing so he heaps discredit on the very enterprise of world-making. "What Plutarch said of the Stoiks, that they spoke more improbabilities than the Poets, may be extended to a great part of Philosophers, who have maintained opinions more absurd than can be found in any of the most Fabulous Poets, or Romantick Writers," all of whom, according to Keill, "only cultivated their own wild imaginations."[19] Mechanist theories of cosmography are, by their very nature, intellectually bankrupt delusions, mere imaginings of the mind. For Keill, the mystery of the Creation and the Flood are just that, mysterious, and it is the height of folly to attempt to circumscribe and explain such events absent the active, miraculous hand of God. It seems obvious to him that the earth we inhabit is "the result of wisdom and counsel; and not to the necessary and essential Laws of motion and gravitation . . . I always wondered at the wild and extravagant fancy of the Philosophers who thought that brute and stupid matter would by it self, without some supreme and intelligent director, fall into a regular and beautiful structure" (44–45).

Keill begins his argument with a historical survey of the folly of both ancient and modern philosophers. His chief target in this critical historiography is, not surprisingly, Rene Descartes, whom he accuses of writing fallacious natural philosophy—not, as vulgarly assumed, with the style and certainty of geometric proofs—but, in fact, in complete contempt of scientific reality.[20] After exposing the scientific failings of Descartes' philosophy,[21] Keill suggests that the problem for Descartes, and for the sect that has grown up around him, is nothing other than overweening ambition and a weakness for illusory theories that seem to explain all phenomena. Though Descartes was "the great Master and deliverer of the Philosophers from the tyranny of Aristotle," ultimately he is "to be blamed for all this, for he has encouraged so very much this presumptuous pride in the Philosophers that they think they understand all the works of Nature" (11–12). Keill specifically decries the way Cartesians like Burnet seize upon false and/or simplistic ideas about mechanism to "make worlds"—that is, to dream up an apparently comprehensive account of all phenomena in nature. Thus we can see for Keill, the question isn't whether this theory or that theory is right in this or that particular. Rather, the very process of "making worlds" is his chief complaint. Although "much more absurd" than Aristotle, "Mons. DesCartes' ingenious hypothesis... his followers pretended, could solve all the phaenomena in nature, by his principles of matter and motion, without the help of attraction and occult qualities. He was the first world-maker this Century produced..." (14).

Why Keill is so intent on debunking Burnet and "world-makers" is clearly beyond a simple interest in correcting erroneous views, or in securing Christianity against deists. Certainly, Keill is at pains to debunk the entire edifice of Cartesian physics upon which Burnet constructed his cosmography: his deriding their failure to account for Newtonian attraction in the foregoing quotation makes that clear, and his text never misses an opportunity to debunk Cartesian doctrines.[22] However, it would seem that Keill's ultimate goal in producing this tract is to debunk the very practice of cosmography as arrogant, impious, and, worse of all, pernicious. With all these "contrivers of deluges" about, "deducing the origination of the Universe from Mechanical principles" have become, according to Keill, something like an imaginative fashion, thanks to Descartes, "who first introduced the *fancy* of making a World."[23]

To direct his argument primarily against Burnet's almost twenty-year-old theory might seem an arbitrary (or at least, a safe) choice, so as to avoid embarrassing more recent theorists like Whiston and Woodward, it would seem that Keill takes aim at Burnet precisely because of the staying power of his cosmography. Keill says that his belated critique of Burnet is warranted because the *Sacred Theory* " has not been so fully refuted as it might have been, nor has anyone shew'd the greatest mistakes in it" (22). As part of this, though, Keill acknowledges that, in defiance of logic, there is something about Burnet's performance that resists easy dismissal:

> There was never any book of Philosophy written with a more lofty and plausible style than it is, the noble and elegant description the Author gives of the subjects he treats shews that he has a great command of

> Language. His Rhetorical expressions may easily captivate any incautious reader, and make him swallow down for truth, what I am apt to think the Author himself, from the sacred character he bears, designed only for a Philosophical Romance. . . . he has acted the part of the Orator much better than he has done that of the Philosopher. (26)

Indeed, Keill acknowledges that despite the *Sacred Theory's* evident scientific failings, "Perhaps many of his Readers will be sorry to be undeceived, for as I believe, never was any Book fuller of Errors and Mistakes in Philosophy, so none ever abounded with more beautiful scenes and surprising Images of Nature" (175). Readers *want* to believe in Burnet's cosmography, such is its imaginative power. And it is that power that even Keill acknowledges might well have a use and value in polite discourse. In concluding his assault on Burnet's theory, Keill says, "I write only to those who might perhaps expect to find a true Philosophy in it. They who read it as an Ingenious Romance will still be pleased with their Entertainment" (175–176).

Cosmography as "Philosophical Romance"

To judge from Burnet's reception in the eighteenth century, polite readers were well pleased with their entertainment for generations. Despite serious religious and scientific scruples, Burnet's *Sacred Theory* remained widely read and much-beloved for its uniquely vivid and pleasurable vision of a world founded in a harmonic synthesis of reason and faith—so much so that one prominent eighteenth-century critic wrote that Burnet "displayed an imagination very nearly equal to Milton."[24] Indeed, the *Sacred Theory's* richly descriptive style and uplifting brand of physicotheology provided a model for the kinds of rational pleasures of the imagination that Joseph Addison—who wrote a Latin poem in praise of the work that introduced all eighteenth-century editions—would later articulate in "The Pleasures of the Imagination," the multipart aesthetic treatise first published in the *Spectator* 411–421, which was profoundly influential in setting up the eighteenth-century standard of taste that found order, beauty and—above all—the hand of divine design in the contemplation of the natural world.[25] As early as *The Spectator* 146, long after Burnet's book was discredited scientifically, Joseph Addison chose *The Sacred Theory of Earth* as a perfect subject for his readers' speculations that day, when, in a high-minded mood, he proposed that his readers seek out "the highest pleasure our minds are capable of enjoying with composure," the reading of "sublime thoughts communicated to us by men of great genius and eloquence."[26] Clearly, in Addison's hands we have moved a great deal beyond the question of whether Burnet's is literally true or not; rather, the book becomes the occasion for enjoying "the highest pleasure," "sublime thoughts." In characteristic fashion, Mr. Spectator exhorts his readers to visualize and imaginatively inhabit his subject, which he praises in the following manner:

> Burnet has, according to the lights of reason and revelation, which seemed to him clearest, traced the steps of Omnipotence. He has with a celestial ambition, as far as it is consistent with humility and devotion, examined the ways of Providence, from the creation to the dissolution of the visible world. How pleasing must have been the speculation, to observe Nature and Providence move together, the physical and moral world march the same pace: to observe paradise and eternal spring the seat of innocence, troubled seasons and angry skies the portion of wickedness and vice.

Here then we can trace the mode of reading that eventually recuperates Burnet's contentious late-seventeenth-century cosmology. Addison and his readers have learned to dislodge Burnet's cosmology from the minute particulars of its natural philosophy or its theological commitments, and instead prize it as a means for producing edifying thoughts about nature; indeed, we might even say it serves as a kind of mechanism for inducing sublime transports of speculation. Addison's Mr. Spectator imagines Burnet himself as a kind of ideal and divinely gifted observer—a Spectator par excellence—whose painstaking synthesis of a coherent intellectual account of the natural history of the earth is figured as a concrete, visual "speculation" of the scenes of earth's mysterious origins. Addison's intention with this essay is, of course, to cultivate piety, but it would seem that his ultimate effect is to cultivate a state of mind in his reader that constructs a kind of spectacle out of the natural world, and which finds astonishment and delight—if not necessarily solid knowledge—in cosmological contemplation.

This kind of natural philosophic pleasure—implicit in Burnet, and made explicit by Addison—suggests the endpoint of the trajectory of late seventeenth-century cosmology, where physicotheology gives way to the cultivation of exquisite states of affective pleasure. The goal is not necessarily the teaching of a specific doctrine, but rather the formation of a mass audience of enthusiasts, who not only can apprehend the essential concepts of natural philosophy, but who have learned a way of speaking about natural philosophy and integrating it into polite culture through the work of imaginative cosmology. The popular audience for science that Burnet helped call into being may not have understood the particulars of Enlightenment natural philosophy, but they knew of it, and they knew what it was good for.

Notes

1. Ernest Tuveson, "Swift and the World Makers," *Journal of the History of Ideas* 11: 1–71. For other classic accounts of the "deluge controversy"

incited by Burnet, see Marjorie Hope Nicolson, *Mountain Gloom and Mountain Glory: The Development of the Aesthetics of the Infinite* (Ithaca: Cornell University Press, 1959) and Joseph Levine, *Doctor Woodward's Shield: History, Science and Satire in Augustan England* (Berkeley: University of California Press, 1977). See also James E. Force, *William Whiston, Honest Newtonian* (Cambridge: Cambridge University Press, 1985).

2. See, particularly, Peter Harrison, *The Bible, Protestantism and the Rise of Natural Science* (Cambridge: Cambridge University Press, 1999); Kevin Kileen, "A Nice and Philosophical Account of the Origin of All Things: Accommodation in Burnet's *Sacred Theory* (1681) and *Paradise Lost*," *Milton Studies* 48 (2007): 113. See also Kenneth J. Howell, *God's Two Books: Copernican Cosmology and Biblical Interpretation in Early Modern Science* (South Bend: University of Notre Dame Press, 2002), and the classic study, Amos Funkenstein, *Theology and the Scientific Imagination from the Middle Ages to the Seventeenth Century* (Princeton: Princeton University Press, 1986).
3. Harrison, *Bible*, 134ff.
4. Kileen, "Accommodation," 113.
5. Harrison, *Bible*, 140.
6. Stephen Jay Gould, *Time's Arrow, Time's Cycle: Myth and Metaphor in the Discovery of Geologic Time* (Cambridge, MA: Harvard University Press, 1987), 42. Burnet's *Sacred Theory* is a touchstone for Gould, displaying overlapping premodern and modern notions of time; In Gould's reading, Burnet's text is animated by an irresolvable contradiction between imagining time as either eternal and cyclical or linear and historical, one that is apparently resolved through the Burnet's storytelling sleight of hand.
7. Thomas Burnet, *The Theory of the Earth: Containing an Account of the Original of the Earth, and of all the General Changes Which it hath already undergone, or is to Undergo, Till the Consummation of all Things. The First Two Books Concerning the Deluge and Concerning Paradise* (London: 1684), Dedication, unsigned leaf.
8. Ibid.
9. John Dryden, *Essay of Dramatic Poesy* in *Dryden: A Critical Edition of the Major Works*, ed. Keith Walker (Oxford: Oxford University Press, 1987), 104.
10. Ibid., 105.
11. Burnet's spectacular rhetoric carries through the whole of his text. To cite just one further example, the first volume concludes with this remarkable flight of rhetoric: "Dividing the World into two parts, Past and Future, we have dispatched the first and far greater part, and we come better half of our way; and we make a stand here, and look both ways, backwards to the Chaos, and the Beginning of the World, and forwards to the End and consummation of all things, though the first be a longer prospect, yet there are as many general Changes and Revolutions in Nature in the remaining part as have already happen'd; and in the Evening of this long Day the Scenes will change faster, and be more bright and illustrious" (326).

12. For the chief theological responses to Burnet's work, see Erasmus Warren, *Geologia: or, A Discourse Concerning the Earth Before the Deluge* (London: 1690); and John Edwards, *Brief Remarks upon the Mr. Whiston's New Theory of the Earth* (London: 1697), and especially *A Discourse Concerning the Authority, Stile and Perfection of the Books of the Old and New Testament* (London: 1693).
13. Burnet, *A short consideration of Mr. Erasmus Warren's Defence of his Exceptions against the Theory of the Earth: in a Letter to a Friend* (London: 1691), 10.
14. Harrison, *The Bible*, Introduction, *passim.*
15. Quoted in Edwards, *Discourse,* 35.
16. Edwards employs a slippery slope argument here, claiming that "if in a literal and historical Sense there was no such thing as that *first Disobedience* of Adam . . . then it would follow that Mankind had no need of a *Savior* and *Redeemer,* then *Christ's* Coming in the Flesh was in vain; then all Christianity falls tot he Ground; then when the Writings of the New Testament speak of *Eve's being deceived* . . . and that in *Adam all died,*" ibid., 37.
17. Charles Blount, *The Oracles of Reason* (London: 1693). In the *Archeologia Philosophicae,* addressing a Latinate audience of churchmen and social elites, Burnet expressed an extreme skepticism of the veracity of Moses, claiming that he was forced to radically simplify and distort his history to suit the debased and vulgar understanding of the Israelites. Blount mounted a sustained defense of Burnet's treatise in the book, as well as translated two chapters from Burnet's text that show that the Bible is a patchwork of fables.
18. By the end of the seventeenth century, the scientific standing of Burnet's *Sacred Theory* was significantly eroded; see particularly John Keill, *An examination of Dr. Burnet's Theory of the Earth, together with some remarks on Mr. Whiston's New Theory of the Earth* (London: 1698). Other major scientific responses to Burnet's *Sacred Theory* include John Woodward, *An Essay Towards a Natural History of the Earth* (London: 1685); William Whiston, *A New Theory of the Earth from its Original to the Consummation of All Things* (London: 1696); John Arbuthnot, *An Examination of Dr. Burnet's Theory of the Earth* (London: 1697); Thomas Robinson, *New Observations on the History of the World of Matter* (London: 1696); John Beaumont, *Considerations on a Book entitl'd The Theory of the Earth, Publisht some Years Since by the Learned Dr. Burnet* (London: 1692–1693); and John Ray, *Three Physico-Theological Discourses* (London: 1693).
19. Keill, *Examination,* 1.
20. Among the features of Burnets theory that Keill debunks are as follows: the theory of the earth's crust solidifying out of an oily scum atop the waters of the interior abyss (Burnet falsely applied the laws of specific gravity) (43); Burnet's notion of the primordial earth as smooth as a billiard ball (mountains would have been needed for weather and water collection) (60–61); Burnet's claim that water appeared on the world before the flood because

of the sun evaporating water out of cracks in the crust (Keill mathematically demonstrates that according to Burnet's model there would have been 5,000 times less water evaporated and 10,000 times less water available to any given area as compared to the present earth) (90); and Burnet's claim that the original earth had a perpendicular axis that created a perpetual spring (Keill shows that unchanging angles of solar radiation would have froze all the earth except the middle latitudes, which would have been scorching (74ff.).

21. For example, Keill notes that Descartes claimed that after God merely created an initial quantum of matter, by its necessary mechanisms even animal life could have come into being automatically. Then he asks how Descartes could pretend to such an outrageous claim, when in the most fundamental part of his theory, of the seven rules for motion that he lays down, only one has proved true (13–14).
22. For example, Keill demonstrations that the Cartesian theory of vortexes cannot account for the celestial motions recorded by Galileo and Kepler (16–17).
23. Keill, *Examination,* 19, my emphasis. Of course, it must be noted that Newton's *Principia* concluded with a section entitled "The System of the World" that outlined a comprehensive account of the action of gravity in the constitution of the solar system. Notably, it was this section alone that was first translated into English. Still, it would be safe to assert that Newton himself, and his early adherents and explicators like Keill—if not the exuberant Newtonian champions of mid-eighteenth century like Henry Pemberton—were careful to circumscribe the limits of what they felt authorized to assert about the reach and comprehensiveness of his theory of gravity. It was in the expressly speculative Queries 28, 30, and 31 appended to the *Optics,* as well as in the circumspect General Scholium appended to the *Principia*—but not in a grand and fully elaborated work of cosmography—that Newton raised the possibility that gravity might be related to other active forces to be found in nature, and thus gesture toward a comprehensive system of nature. "A few things could now be added concerning a very subtle spirit pervading gross bodies and lying hidden in them; by its fore and actions," Newton writes in the General Scholium, going on to note the phenomena of electricity, fluid cohesion, the emission of heat by light, and nervous stimulation, among others, "But these things cannot be explained in a few words; furthermore, there is not a sufficient number of experiments to determine and demonstrate accurately the laws governing the actions of this spirit." Isaac Newton, *Philosophical Writings,* ed. Andrew Janiak (Cambridge: Cambridge University Press, 2004), 92.
24. Joseph Warton, *Essay on the Genius and Writings of Pope* (London: 1806), 1:260. Quoted in Nicolson, *Mountain Gloom,* 193.
25. Addison's essays on the "Pleasures of the Imagination" was originally written as a single treatise, but it was first published in parts in *Spectator* 411–421 (June 21, 1712–July 3, 1712). For a thorough accounting of the impact of Burnet's *Sacred Theory* on eighteenth-century poets and essayists, see

Nicolson, *Mountain Gloom*, Ch. 6, *passim*. Addison heaped high praise on the work in *Spectator* 164 (August 17, 1711), for example, while as late as 1768, the Burnet controversy was versified in Richard Jago's *Edge-Hill* (London: 1768).

26. *Spectator* 146 (August 17, 1711), in Richard Steele and Joseph Addison, *The Spectator with Notes and a General Index Complete in Two Volumes* (New York: Samuel Marks, 1826), 1:189.

CHAPTER 7

The Jesuit Mission to Ethiopia (1555–1634) and the Death of Prester John

Matteo Salvadore

Introduction

"There is only one Catholic Church in the world and it can only be one under the Roman Pontiff and not under that of Alexandria."[1] Thus spoke Ignatius of Loyola in 1555, addressing the Ethiopian Emperor Galawdéwos in relation to the Ethiopian Church's tradition of dependence on the Egyptian Coptic Church. Since its beginnings, with few exceptions, the head of the Ethiopian Church, known as *abun* or metropolitan, was an Egyptian cleric appointed by the patriarch of Alexandria. In Loyola's eyes, the dependence on Alexandria was one among many "errors and abuses" making Ethiopian faith an aberration and requiring the attention of Jesuit proselytism.

In the midst of the era of Catholic renewal, the Society of Jesus embarked on a project that had profound consequences for Ethiopian-European relations and marked a watershed for Europe's idea of Ethiopia. In 1555 the Society dispatched the first party of missionaries to the kingdom with the objective of establishing a *province* under the jurisdiction of the Portuguese *assistancy.*[2] In the Jesuit imagination, Ethiopia would have welcomed its *reduction*[3] to Catholicism and the authority of the appointed *provincial* as patriarch of a new Catholic Ethiopia. Though Loyola's plan failed and was inconsequential for the long-term trajectory of Ethiopian Christianity, the mission deeply affected the unfolding of the Ethio-European encounter that Ethiopian pilgrims had spearheaded during the Renaissance.[4]

This chapter examines the consequences of the Jesuit mission to Ethiopia (1555–1634) on early modern European world-building: starting in the late fourteenth century, Europeans had come to imagine Ethiopia as one of the most alluring destinations in the Orient. European elites identified Ethiopia with the distant Christian land of *Prester John*, the pious king capable of rescuing Christianity from the looming Islamic threat: Italian and Portuguese courts hosted Ethiopian pilgrims and diplomats, whom they recognized as peers.[5] However, once at the court of Emperor Galawdéwos (1540–1559),[6] the first Jesuit missionaries chose *accomodatio* over *impositio,* disregarding the cautionary tale of two fathers who visited the emperor in 1555.

One of the two Jesuit emissaries, Gonçalo Rodrigues, returned to Goa disappointed about the encounter and vented his frustration in a letter to his superiors. With uncanny insight, he warned his superiors that "the notables of the empire would prefer to be subjects to Muslim rule rather than replace their customs with ours."[7] Despite Rodrigues's warning of a coming storm, the Jesuit party headed by Andrés de Oviedo (1517–1577) reached Galawdéwos's court in 1557 and immediately demanded the unconditional acknowledgment of João Nunes Barreto (1510–1562)—the first appointed *provincial* of Ethiopia—as patriarch of all Ethiopian Christians.[8] His suggestion was met with a marked lack of enthusiasm that reflected the vast disconnect between Ethiopian and European ideas about Ethiopia and Ethiopian Christianity. Evidence suggests that through Jesuit mediation, the European imagination reinvented Ethiopia: the Society transformed the kingdom of *Prester John* into a land of mission, articulated a discourse of conquest and attempted to enforce Eurocentric standards. By treating Christian Ethiopians as nonbelievers, the Jesuits fractured Ethio-European commonality and shaped a new discourse of otherness: no longer were Ethiopians Christian brothers to be respected, but rather heretics to be converted.

The nature of the Ethio-European encounter before the coming of the Society is fundamental to understanding how the Society's rejection of Egyptian Christianity changed their relationship. Ignatius of Loyola's instructions to João Nunes Barreto provide clear insight into the substance of the Jesuit mission in Ethiopia and reflect European ideas about that nation and its faith. Finally, I review the history of the Jesuit sojourn in Ethiopia (1554–1633) and its cultural by-products. The collection of historical and protoethnographic accounts the fathers produced embodied the long-lasting legacy of a process of material and spiritual conquest that failed on Ethiopian soil but successfully changed discourse about Ethiopia among Europe's learned circles.

The Image of *Prester John* before the Luso-Ethiopian Encounter

Starting in the fourteenth century, Ethiopian travelers ventured to Mediterranean Europe for a variety of reasons. Clerics undertook pilgrimages to Rome, Padua, and Santiago de Compostela out of devotion and, as fellow Christians, found the enthusiastic support of Europe's ecclesiastical hierarchies. The most fortunate received letters guaranteeing food and lodging in Europe, for which their benefactors were recognized by the Church with letters of indulgence.[9] Learned Ethiopians and trusted *färäng*[10] were dispatched as representatives of either the Ethiopian Church or the Ethiopian emperor. Regardless of the reason for their visits, all across Mediterranean Europe Ethiopians were welcomed as peers whose black skin and alien culture remained inconsequential in light of their faith. The specter of *Prester John* held such sway in European Catholic circles that even those with scant credentials willing to claim ties to him were welcomed as special guests in Europe's most prestigious courts.[11]

The myth traces its origins to the Middle Ages, in particular to 1165 when Byzantine emperor Manuel I Komnenos (1143–1180) received a long letter through which a self-declared *Prester John* sought alliances with his European peers.[12] Undisputedly a forgery, the letter was an inventive collection of geographical and pseudoethnographical information from a variety of classical and medieval works that read like a compendium of classical knowledge injected with fragments of contemporaneous information obtained from European travelers to the Far East.[13] *Prester John* represented the archetype of the perfect Christian sovereign: his kingdom was rendered as the counterpoint to a Europe and Middle East that had been ravaged by war and as the deus ex machina capable of resolving the perennial confrontation between West and East, Christianity and Islam.[14]

Until the early thirteenth century, the *Prester* was thought to be located somewhere in the Far East and possibly be the Mongol Empire. By the end of that century, however, European encounters with the Mongols prompted the relocation of the imaginary sovereign to Ethiopia. Among these were the accounts of Franciscan pilgrims such as Giovanni da Pian del Carpine (1180–1252) and William of Rubruck (1120–1293), who traveled throughout the Mongol Empire in the 1240s and 1250s and finally dispelled the myth of an almighty Christian king in Asia, and the thirteenth-century clashes between the Mongol Empire and European military forces.[15] Europeans were left searching for a new place for their exotic Christian king, and Africa, with its emergence onto the world stage, became a logical site for him. This shift was certainly not unidimensional. Between the fourteenth and fifteenth centuries Ethiopians successfully appropriated the myth of the Christian king of the East, transfiguring their own country into the land of the devout *Prester*.

In more general terms the Africanization of *Prester John* coincides with the emergence of what Janet Abu-Lughod described as the emergence of "a complex and prosperous predecessor [to the modern world economy], a system of world trade and even 'cultural' exchange that at its peak toward the end of the thirteenth century, was integrating a very large number of advanced societies stretching between the extremes of northwestern Europe and China."[16] The creation of a Euroasian world economy and the coming of a merchant class and long distance trade contributed raw material to the process of world-building by feeding firsthand observation, hearsay and myth to Europe's planetary consciousness.[17]

On July 16, 1402, a group of Ethiopians reached Venice and was greeted as ambassadors of "excellens dominus Prestozane, dominus partium Indie"[18] Between the late Middle Ages and the early modern period Ethiopians became key agents in the world-building process that made their own country available to Europe's reading public at a time when only a handful of Europeans had reached the Ethiopian highlands.[19] They shared geographical knowledge with prominent cartographers and led their interlocutors to believe that Ethiopia was indeed the kingdom of the fabled *Prester.*[20] The visitors demonstrated a good understanding of Southern Europe's geopolitical chessboard: they first focused their diplomatic activity toward Venice in the early fifteenth century, to then shift their attention to Rome—which they soon recognized as the source of organized Christianity—and by the mid-fifteenth century to the Kingdom of Aragon, which was emerging as a key player in the Mediterranean basin.

Despite several initiatives on both sides of the encounter, fifteenth-century interactions remained quite inconsequential in the short period, as no alliance came to fruition; however, contacts between Europeans and Ethiopians throughout the Mediterranean basin contributed to the production and diffusion of knowledge about Ethiopia and fueled the myth of the Ethiopian *Prester* believed capable of controlling the waters of the Nile and of blackmailing the Muslim world. Niccolò da Poggibonsi, a Franciscan friar who visited Jerusalem between 1345 and 1350 and once there became acquainted with its Ethiopian community, so characterized Ethiopians and their sovereign:

> This generation [of Ethiopians] loves us Christian Francs above all and it would be happy to join us; but the Sultan of Babylon never let any Latin head toward them pass. But these [people] of Tiopia can pass and come to Egypt and to the Land of Mission [Jerusalem] without paying tribute to the Sultan; and they can wear the cross in the open throughout Saracinia, they even enter the Holy Sepulcher without paying tribute; nobody else has this privilege apart from those from Tiopia; this is what the Sultan does: of course it does it for fear, because the Lord of Tiopia is the greatest Lord in the world.[21]

While politically inconsequential, the fifteenth-century encounter and the resulting production of knowledge stimulated further interest in the distant Christian land and in particular informed Portuguese exploration in the aftermath of the *Reconquista.*

The Discourse of Sameness in the Luso-Portuguese Encounter

One of the primary purposes behind the Portuguese circumnavigation of Africa, started under the leadership of Prince Henry *the Navigator* (1394–1460) in the early fourteenth century, was to identify a path to Ethiopia, and secure *Prester John*'s support against the growing Islamic threat.[22] Several Portuguese parties attempted to reach Ethiopia from the *Sinus Aethiopicus* in West Africa, the Swahili Coast, and through the beaten path that connected Ethiopia to Europe through Jerusalem and Cairo.[23] With the partial exception of a Portuguese envoy who reached the Ethiopian court in 1491 but failed to inform Lisbon of his whereabouts until a second mission reached the country, all these attempts resulted in failures. It would only be in 1508 that Portuguese representatives succeeded in reaching the *Prester's* court. Shortly thereafter the Ethiopian regent Eleni (1431–1522) dispatched a representative to the Portuguese court and officially opened the way to formal diplomatic relations between the two countries, an act that would be reciprocated with the dispatch of the Portuguese embassy to Ethiopia in 1521.

These exchanges resulted in an unprecedented production of knowledge about Ethiopia and in the publication of the first European eyewitness accounts about the country. While more or less reliable references to Ethiopia had abounded in a variety of publications throughout the fifteenth century, the new accounts offered information previously unavailable. The two most emblematic examples of this new phase of the encounter are those of two Ethiopian representatives, Matewos and Zagā Za'āb, who reached Lisbon respectively in 1514 and 1527. They spent considerable time with the clerical and secular elites in the Portuguese capital and authored accounts about Ethiopian Christianity that were published under the supervision of Portugal's first Ethiopianist, Damião de Góis (1502–1574). After having acted as informants to a variety of European intellectuals throughout the fifteenth century, Ethiopians contributed more directly to European world-building: their contributions represented both opportunity and challenge for the Portuguese intellectual elites who understood the importance of commanding the narrative of their experience overseas.[24]

By the same token, individuals involved in the Portuguese embassy who left for Ethiopia in 1515 produced more than one account of the tortuous path that eventually brought them to the Ethiopian highlands. Andrea Corsali, a papal emissary and a correspondent of the Medici family, recorded his experience on the Red Sea in two letters that were published in 1516 and later included in Giovanni Battista Ramusio's (1485–1557) collection of travel narratives—*Navigationi et Viaggi*.[25] Diogo Lopes de Sequeira and Pero Gomes Teixeira, respectively governor and auditor of the *Estado da Índia*, are the most likely authors of the letters. They landed on the Ethiopian coast with the embassy and spent some time in the area before returning to Goa. While there, they authored letters that were published in 1521 in Lisbon under the title *Carta das Novas que Vieram a el Rey Nosso Senhor do Descobrimento do Preste Joham*.[26] Finally, the embassy's chaplain, Francisco Álvares (1465–1540), authored the famous *Verdadera Informaçam das terras do Preste Joam das Indias*, which was published posthumously in 1540.[27]

Álvares was the first author to publish an extensive eyewitness account of what was still a little known country. He told the story of his journey to the Red Sea, his first encounter with Ethiopian authorities in the vicinity of Massawa, his journey to the Ethiopian highlands and his stay at Lebna Dengel's court. *Verdadera* was a groundbreaking work, as confirmed by the multiple translations and editions, whose existence speaks to the complex network of intellectuals that spanned from Lisbon to Rome and Antwerp. Furthermore, *Verdadera* was prominently featured in the collections of Giovanni Battista Ramusio, Richard Hakluyt, and Samuel Purchas, authors responsible for introducing travel narratives to the European reading public and, in the process, laying the ideological foundations of European expansion.[28]

Verdadera best epitomizes the discourse of sameness that defined the Ethio-European encounter before the coming of the Jesuits. Throughout his account Álvares remained faithful to a relativistic and ecumenical stance that after having defined the encounter since its inception, would soon give way to its antithesis, the Jesuit imposition of Catholicism by all means necessary. In the opening pages of his work, Álvares tells the reader that

> I have decided to write down everything that happened on this journey and the countries where we were, and their characteristics, customs and practices, which we found there, and how they conform to Christianity, not ensuring or approving their customs and usages, but leaving everything to my readers to praise, amend or correct as shall seem best to them.[29]

In fact, *Verdadera* surprises for the sheer absence of any ethnocentric discourse: what Álvares found different, he approached descriptively rather than prescriptively. In his discussion of a number of practices, such as marriage, divorce, circumcision, and baptism, that we know struck him as peculiar, he refrained from dispensing criticism or verbalizing repulsion for Ethiopia's traditions. The same practices that less than a century later Jesuit authors would present as the ultimate proof of Ethiopian heresy and otherness seemed only interesting to Álvares. Furthermore, skin color is presented as largely inconsequential: he praises and criticizes his interlocutors on the basis of their social and economic status and religious persuasion. Overall, Álvares's discourse on Ethiopia and Ethiopians seems that of a detached observer.

On October 20, 1521, Álvares reached Lebna Dengel's court: the chaplain describes the sovereign as very much focused on the issue of importing European technology and constantly asking the Portuguese ambassador Lima about weaponry.[30] In one of the meetings Lebna Dengel asked "if any of us [Álvares and the Portuguese] could make powder. [...] He said that sulphur could be found in his kingdoms, even that there were craftsmen to make saltpeter. All his armies lacked was the use of artillery and someone to teach them to work it, because he could marshal innumerable carabineers with whom he would subdue all the neighboring Moorish kings."[31] The Ethiopian quest for superior Western technology would be met by Jesuit willingness to share practical know-how for the purpose of presenting Catholicism as progress: in fact, technological transfer was the most successful component of the Jesuit strategy in Ethiopia, a country that was in fact more interested in progress than in Catholic orthodoxy.

The decades following Lima's mission saw both the emergence of an existential threat to Ethiopia and the climax of the Ethio-European encounter. Thanks to a two-century long process of world-building complicated by contingencies, misconceptions, myths, and political calculations, Ethiopian and Portuguese elites regarded each other as peers and allies against Islam. By 1529 Lebna Dengel faced an invasion of unprecedented proportion as the army of the Sultanate of Adal, lead by Ahmad ibn Ibrihim al-Ghazi (1506–1543), marched onto the highlands. Thanks to the numerous cases of interactions over the preceding decades and the intercession of a member of the Portuguese mission who had stayed behind in Ethiopia, in 1543 Portugal intervened in the conflict: with the help of a contingent of four hundred soldiers the Adali were defeated. The victory was the culmination of a long process of reciprocal discovery between two polities on opposite sides of the Mediterranean basin who understood each other as belonging to the same Christian world.[32]

By 1546 when Ignatius of Loyola redacted his instructions for the Jesuit mission, the protagonists of the Ethio-European encounter had engaged in diplomatic relations, a military alliance, technological transfer, extensive production of secular and theological knowledge and, finally, had welcomed each other's diasporas. While the Portuguese soldiers who contributed to defeating the Adali were settling in Tigray, several Ethiopian monks had found a new home in Rome in the monastery of *Santo Stefano degli Abissini*. In both cases the foreign visitors were supported and befriended by their hosts. On the one hand, the Portuguese community in Tigray was granted land rights, supported by the local elites, allowed to marry local women and raise children in the Catholic tradition—clear signs of a long-established tradition of tolerance toward Christians of different rites. On the other hand, the Ethiopian community in Rome enjoyed the generous support of the Papacy and prominent Roman families in what seemed to be a clear sign of long-term commitment to an ecumenical dialogue.[33]

Rome, and to a lesser degree Lisbon, had become the foremost center of knowledge production in Catholic Europe, and authors from both sides of the encounter had turned the first half of the sixteenth century into the age of the Ethio-European text. Printing presses had been replenishing the shelves of the Ethiopianist library with works that for the most part represented a joint transcultural effort. These developments were taking place against the backdrop of the Counter-Reformation, however, which would soon have paramount consequences for the encounter and the future of Catholic ecumenicalism.

An early warning of the coming storm could be appreciated in Lisbon during the sojourn of the Ethiopian erudite Zagā Za'āb at Joao III's court between 1527 and 1533. After suffering the scorn of court theologians and being denied communion, the Ethiopian decided to author *The Faith of the Ethiopians*, first published in 1540.[34] The account was a thorough refutation of the accusation of heresy, affirmed the theological soundness of Ethiopia's Christian traditions, and denounced the narrow-mindedness of the Catholic Church. It also speaks to the broader issue of Catholic relations with other Christians and suggests an accommodating attitude vis-à-vis the reformed churches in Europe. Zagā Za'āb can in fact be considered an ante-litteram critic of the Counter-Reformation: as such, he found the support of Damião de Góis, Portuguese intellectual and foremost court historian trained in the Erasmian tradition of Christian humanism.[35]

For a peculiar set of circumstances, among which his presence at Manuel's court at the time of the first Ethiopian embassy to Portugal in 1518, Góis became involved in the theological debate about Ethiopian Christianity.[36] Apart from befriending Zagā Za'āb and championing the Ethiopian cause, the scholar corresponded with several personalities in the Lutheran camp and seemed reluctant to accept the Papal excommunication of the Reformers. In fact, Góis paid dearly for his openness toward Christians outside the boundaries of the Catholic Church: the first setback for his standing as a trustworthy Catholic came in 1541, when the Grand Inquisitor Cardinal Henrique informed Góis that *Faith* had been banned.[37] In 1545 he was dismissed from the Portuguese court, and then two failed attempts were made to bring him in front of the inquisition.

A central figure behind Góis's misfortunes was Simão Rodrigues, an influential Portugese courtier who was also one of the first six Jesuits. He had crossed paths with Góis decades earlier and remained unimpressed with his openness toward Ethiopians and Protestants.[38] In 1569 Góis was imprisoned and later tried for heresy. Found guilty thanks to Rodrigues's deposition, he was granted reconciliation, deprived of his property and confined to a monastery. He died in 1574, one among many of the illustrious victims of Catholic renewal. Together with Zagā Za'āb's mistreatment, his changing fortune foretold the story of the growing Jesuit influence in the Christian world and of the coming paradigm shift in Ethio-European relations.[39] In fact, by the early seventeenth century, the Portuguese community in Tigray had disappeared; most Ethiopian and mixed-race Catholics had been either forced to convert or relocate. The monastery of *Santo Stefano degli Abissini* lost most of its Ethiopian guests, and the most prominent European scholar to collaborate with an Ethiopian had been excommunicated.

Little is known about the Ethiopian clerics in Rome during the era of the Jesuit mission, but fragments of information suggest a change of disposition of the Catholic Church toward the Ethiopian vanguard of the encounter. In the 1580s Gregory XIII (1572–1585) granted the Ethiopians in *Santo Stefano* increased material support from the curia—apparently evidence of his desire to invest in the ecumenical relation—but in the same ordinance stated that the Ethiopians "must be received in the College of Propaganda Fide, where they will have all comforts to learn the Catholic faith, which is the only reason to provide for them and feed them."[40] In fact, we know that in 1596 the Jesuits dispatched an Ethiopian monk by the name of Takla Maryam to Rome to lobby for more support. Before dispatching him, the Jesuits decided to reordain him, marking a clear divide between what they considered the Catholic world on one hand and heresy on the other.[41] Furthermore, in 1628 *Santo Stefano* lost its collection of Ethiopian manuscripts to the Vatican Library—an occurrence that mirrored the growing Catholic desire to control and manipulate the production and regulation of knowledge related to Ethiopia. Starting in the second half of the sixteenth century, *Santo Stefano* also saw a continuous decline in the number of hosted Ethiopians and eventually became the residence of Ethiopian converts.[42]

The ecumenical and welcoming *Santo Stefano*—where Ethiopian and European clerics had been engaging each other in discussions of liturgical, theological, and geographical knowledge had turned into a college for Ethiopians willing to accept Catholic supremacy.[43] What erased two centuries of Ethiopian attempts to attract European interest for their country, several centuries of European amazement in the face of distant Christians, and one hundred and fifty years of collaborative relations hinged on a shared Christian identity? In the second half of the sixteenth century the process of reciprocal discovery that had defined the Ethio-European encounter was obliterated by new agents of global change. The pillars of the encounter—the conflation of medieval legend with early modern world-building and the combination of Ethiopian cosmopolitanism with European enthusiastic reception—suddenly became anachronistic in light of the new religious paradigms of Catholic renewal. Counter-Reformation Catholicism replaced the ecumenical idea of Ethiopians and Europeans sharing the same Christian identity with the notion of a contraposition between Catholics on one side and heretics on the other.

For the first time since the Council of Chalcedon of 451, the Catholic Church chose aggressive proselytism over ecumenical dialogue. At that

Council, a theological dispute relative to the true nature of Christ had created a paradigmatic divide between the Catholic and most of the Eastern Churches on one side and a minority of Eastern Churches on the other, among which were the Egyptian and the Ethiopian Church. However, the history of the Ethio-European encounter shows that in spite of the schism, the Catholic Church sought cordial relations with the Ethiopian Church throughout the fourteenth and fifteenth centuries. While maintaining a critical stance toward Ethiopian practices, the Catholic Church had welcomed Ethiopian clerics in Rome, embraced dialogue and generally considered Ethiopians part of the Christian community. In the age of Catholic renewal the stance of the papacy underwent a radical change, and a newly founded religious order targeted Ethiopia as a land of mission: its soldiers conquered, however provisionally, the Ethiopian highlands in a matter of decades.[44]

Loyola's Instructions

The seeds of the Ethiopian chapter of the Christian wars of religion are located in the founding document of the Jesuit mission to Ethiopia, the *Requerdos que podran ajudar para la reduction de los reynos del Preste Juan a la union de la yglesia y religio catholica*. Loyola authored the *Instructions* in 1553 to offer guidance to the newly appointed Catholic patriarch of Ethiopia, João Nunes Barreto (1510–1562). Loyola demonstrates a certain degree of familiarity with Ethiopia and its history: we know he consulted Álvares's account and that he was introduced to the Ethiopian community in Rome. In particular he corresponded with Tasfā Seyon (n.d.–1550), prominent Ethiopian monk at the monastery of *Santo Stefano degli Abissini* in Rome.[45] Cognizant of Ethiopia's hierarchical social structure, Loyola advised the provincial to first seek converts among the elites "as the result of the enterprise we aim at depends humanly speaking on Prester John or King of Ethiopia and subordinately on the people."[46] For Loyola, the *Prester* and the nobility were the primary targets of the mission, whose agenda was to be implemented slowly and prudently: "go with gentleness and without resorting to violence towards those souls."[47] As we will see, the fathers who remained faithful to the *Instructions* proved to be quite successful in their endeavor; but those who chose to ignore Loyola's advice had particularly negative consequences for the mission and ultimately determined its failure.

Unlike his ideas about how to convert all of Ethiopia through its elite, much of Loyala's approach owed more to his imagined version of Ethiopia than accurate textual representations. In 1555 Loyola wrote to Galawdéwos that the Egyptian patriarch had no authority: "The patriarch who is in Alexandria or in Cairo is schismatic and separate from the Apostolic Siege and the Pontiff, who is the head of the whole body of the Church, does not recognize to him either grace or authority [. . .]."[48] Loyola was oblivious to the country's history and the deep roots of its complex religious tradition of voluntary dependence on the Egyptian Church.[49] One of the documents at the origin of the misunderstanding was Lebna Dengel's 1533 letter to Clement VII. Like many other contemporaries and later scholars of Ethiopia, Loyola considered it to be an abjuration of Ethiopian ecclesiastical independence and a statement of submission to the Catholic Church.[50] For this reason he affirmed the importance of "mentioning [to the emperor] the obedience that David, father of the current king, sent to the Holy See."[51] However, Lebna Dengel's letter was only an expression of interest for a closer association and, if properly read in the context of the encounter, it cannot be interpreted otherwise.

Loyola discounted the importance of the Ethiopian Church's ties to Alexandria, which might have led to the confusion surrounding João Bermudes's mission to Europe at the time of the Adali-Ethiopian War. Bermudes, who originally reached Ethiopia as the barber and physician of Lima's mission, traveled to Rome and Lisbon in 1535 to obtain military support for Lebna Dengel. Once in Rome he introduced himself as patriarch of Ethiopia, claiming to have been appointed by the emperor himself.[52] While a number of historians have demonstrated the disingenuous nature of Bermudes's actions and the preposterous nature of his claim, the event most likely had important repercussions on Loyola's image of Ethiopia and understanding of its Church. Loyola was familiar with Bermudes's presumed appointment as *Abun*,[53] and it is possible that regardless of the Ethiopian's opinion on the matter, he regarded it as evidence of the emperor's availability to entrust a *färäng* with the patriarchate. Loyola could have reached the conclusion that there was no reason for the Ethiopians to reject a learned and properly anointed Jesuit as patriarch after having accepted a self-appointed and little learned barber. Loyola's confidence led to the institution of a "Catholic Patriarch," established in 1554 by Julius III's bull *Cum nos super*.[54] As it has been recently noted elsewhere, the decision to create a patriarch and its ascription to the Jesuit order was a peculiar one that had very negative effects on Ethiopia's secular and religious elites, which had never sanctioned Bermudes's self-appointment as patriarch.[55]

The *Instructions* also show that Loyola was aware of two Ethiopian prophesies that had made their way to Europe. One, which seems the mirror image of the myth of *Prester John*, originated in the sacred texts of the Ethiopian religious tradition and related that a *Frank King* would come to the rescue of a threatened Ethiopian Empire.[56] The other prophecy adumbrated the union between the Ethiopian Church and Rome, which was supposed to be completed after the death of the hundredth patriarch.[57] Loyola, as a pragmatic negotiator, included multiple references to these prophecies in the *Instructions* and presented them as possible talking points for the missionaries' negotiation with the Ethiopian elites.

One of the strategies Loyola pursued was to build on the Ethiopian elites' eagerness to establish a partnership with Western countries for the purpose of acquiring technology and know-how. The Ethiopian awareness of being materially backward and a relentless quest for foreign technology is in fact a leitmotiv that runs through Ethio-European relations throughout the fifteenth and sixteenth centuries:[58] to the Ethiopian elites, Europeans were not only brothers in faith, but purveyors of technology—an expectation that would indeed be met by the Jesuits. Loyola was aware of the desire and suggested using the technological incentive as a tool for conversion. He suggested the patriarch make clear to the Ethiopian emperor that

> he will obtain true and real union and friendship with the Christian princes, when they will all uniformly have the same religion and that then he would see dispatched masters of all trades that he desires [. . .] We will see and will refer to His Majesty in Portugal whether it will be appropriate to dispatch with them [missionaries] some ingenious men to teach them the way to make bridges, for the purpose of crossing rivers and build and cultivate the land and fish and more, and also some doctors or surgeon, so that it will be manifest to these people that even any material good comes with religion.[59]

Loyola believed that by presenting progress as the other side of conversion the mission's chances of success would have increased exponentially, and rightly so as Pedro Páez's (1603–1622) success demonstrated. Of all the patriarchs dispatched to Ethiopia, Páez was undeniably the most effective in attracting the Ethiopian elites toward Catholicism.

The lure of technological progress was linked to a more violent conversion strategy before the Jesuits established a mission in Ethiopia. In 1555 Gonçalo Rodrigues reached Galawdéwos's court to discuss the arrival of a Catholic patriarch and to persuade the emperor to accept conversion. When presented with the compendium of "errors" that the Jesuit had compiled to make his case for conversion, the emperor responded with his *Confessio Fide*.[60] A frustrated Rodrigues returned to Goa and convinced the designated patriarch Barreto to plea the *Estado*'s authorities for a military escort.[61] In 1563 another epistolary exchange internal to the Society refers to the opinion of another father who reportedly suggested "to envoy to the kingdom [of Ethiopia] numerous Portuguese soldiers who could serve not only the purpose of driving away the Turks and help placing on the throne the nephew of the emperor who surged against him, but also introducing the Catholic faith."[62] In 1567 Oviedo wrote letters to Pope Pious V and Sebastian of Portugal (1557–1578), lamenting the Ottoman embargo and begging to organize a military expedition for the purpose of defending Catholics from both infidels and heretics and arguing that "with your coming we would hold great hope for the reduction of these lands to the union with the Catholic faith and the conversion of heathens."[63] In 1583 Francisco Lopes, who we saw would later become the very last-standing Jesuit on Ethiopian soil, wrote to his superiors in Goa arguing that a military intervention was the only action that would lead to *Prester John*'s conversion.[64]

While active military engagement was never adopted, the Jesuit influence in the late sixteenth and seventeenth centuries considerably expanded. Under Páez's leadership the Jesuits returned to court life and emerged from their limited pastoral presence in Fremona to active proselytism across the empire. In the twenty years between his arrival in Fremona and his death at Susenyos's (1606–1632) court in 1622, the number of fathers in the mission increased from the handful under Oviedo's leadership to more than twenty. By the same token the mission's territorial presence went from only three residences in Tigray to seven residences spread to the four corners of the country.[65] In fact, the expansion of the Catholic presence in Ethiopia, far from being a simple matter of converting elites and commoners, entailed establishing a Catholic space that was increasingly expanded at the expense of Ethiopian Christianity.[66] Páez and his successors were indeed responsible for dotting the Ethiopian highlands with masonry works: the construction of Catholic churches, shelters for the Jesuit fathers and schools for the first generation of Ethiopian Catholics, was an integral part of the conversion process. Páez seems to have been directly involved in a number of projects: the *Chronicle of Susenyons* refers to him as a "master mason," a title that speaks volumes about his perception as a purveyor of technical knowledge.[67]

The patriarch-mason started a tradition that continued in the years to come and saw the Jesuits introduce lime-based construction in the country. Soon after Paez's death, an Indian secular priest employed as Afonso Méndes's chaplain, Manuel Magro, landed in Ethiopia. Upon observing the existence of suitable raw material similar to that he was familiar with in Goa, he proposed to Susenyos the use of lime in construction—a technique that seems to have been a novelty for the country.[68] The historical association of lime-based construction with the Jesuit era—and its perception as a European-imported technology—ran so deep in the Ethiopian imagination that its memory survived the expulsion. In 1887 Augusto Salimbeni—whom the emperor entrusted with the construction of a bridge over the Abbay River—was told that the last ones to embark in a similar project had been some *färäng* several generations before, and that no similar attempt had been made in recent times.[69] In what can be regarded as the ultimate vindication of Loyola's call to present Catholicism as progress, the anecdote shows that two hundred and fifty years after the expulsion, the memory of Portuguese masonry skills and the image of *färäng* as purveyors of knowledge survived.

Apart from the exploitation of the technological gap, the Jesuits attempted and to a certain degree succeeded, at joining the Ethiopian court, befriending emperors as well as regional Ethiopian lords willing to support the *reduction* for personal gain.[70] In the *Instructions* Loyola lays out the details of the process, explaining how to win the favor of the elites: "as far as the abuses, first convince little by little the Prester and those with more authority and then, without noise, once the latter are well-disposed, call a gathering."[71] Once the political and religious elite had been drawn to Catholicism, Loyola believed it was necessary to co-opt the local clergy. To this purpose, he suggested rewarding its most receptive members: "see if it is possible to remunerate good priests through the assignment of abbeys and annuities available to the patriarch."[72]

The Ethiopian youth was also to be considered a primary target for conversion efforts. The education of the younger generations in the Catholic and Latin traditions was seen as a necessary step to create a generational divide between young, eager Catholics and their reluctant fathers. To this purpose Loyola suggested establishing schools and planning for the relocation of promising pupils abroad: "[it would be appropriate] that the Prester sent many of the best ones [pupils] outside his kingdom, to make a college in Goa and, if appropriate, another in Coimbra and one in Rome."[73]

While the tectonic shifts of the Reformation and Counter-Reformation were redesigning Europe's political geography, strong reverberations reached the Christian highlands and destroyed the delicate equilibrium of the encounter between the Jesuits and Ethiopia. They revealed the underlying discourse of conquest that was inherent in Loyola's instructions to convert the Ethiopians and reflects the increasing disconnect between European Catholicism and Ethiopian Christianity.

The Jesuit Experience on the Ethiopian Highlands

In 1557 the first five Jesuit fathers reached Ethiopia under the leadership of Andrés de Oviedo (1562–1577); they had been dispatched to prepare the visit of the anointed patriarch, João Nunes Barreto. Soon after his arrival at the royal court, Emperor Galawdéwos denied Oviedo's claim to accept Barreto as patriarch of Ethiopia: in retaliation Oviedo barred Catholics from associating themselves with Ethiopian Christians under threat of excommunication with a proclamation dated February 1559.[74] This first Jesuit mission advanced ill-advised requests: first, it attempted to replace a millennium-old tradition overnight as if Ethiopia's ties to Egypt were inconsequential and could be rescinded with no consequences for the geopolitical landscape of the Horn.[75] Second, the mission had been assigned, in a clear sign of disrespect for the emperor, to subordinates of the appointed Patriarch. Finally, by addressing the *negusa nagast* and ignoring the Ethiopian Church's hierarchies the Jesuits mistakenly followed in Loyola's footsteps in considering the *Prester* as the ultimate depositary of religious authority when instead more often than not the upper echelons of the Ethiopian Church in alliance with regional nobles, played the role of king-makers.

On the Ethiopian front, Galawdéwos died in March of the same year and his successor Ménās (1559–1563) demonstrated less patience with the missionaries. Once enthroned, Ménās revoked any remnant of religious liberty by forbidding Ethiopian Catholics to partake in rituals administered by the Jesuit fathers and by forbidding the latter the preach in public.[76] Furthermore, Oviedo was banished from his court: the Jesuits repaired to Fremona in Tigray, under the protection of *Bahr Negus* Yeśhāq (?–1578),[77] a Catholic sympathizer and leading voice of dissent toward the emperor. Once in Tigray, the Jesuits attended to the Catholic community composed of the surviving soldiers of Gama's expedition, their Ethiopian spouses, and their children raised in the Catholic tradition: for years, until the Ottoman blockade of the Red Sea, the missionaries periodically received alms from the Portuguese.[78]

The rest of the century saw Ethiopia torn apart by internecine warfare between sectors of the nobility that under Yeśhāq's leadership sought more religious freedom and regional autonomy. The forces that coalesced around the emperor and the *Bahr Negus* had very different ideas about the balance of power within the *Church and State* complex, the kernel of the Ethiopian sociopolitical system, as well as differing theological positions.[79] This divide challenged Ethiopia's political and religious history, which since the late thirteenth century had comprised a feudal structure where political and religious power depended upon each other for survival. On the one hand were the Ethiopian emperors, generally committed to territorial expansion and incorporation of the conquered population into Ethiopian society for the purpose of increasing the tribute-paying peasant class. On the other was the Ethiopian Church, which was eager to expand its presence beyond the northern highlands of Ethiopia into territories historically inhabited either by Muslims or followers of ethnic religions. Starting in the late thirteenth century, the marching order of the Empire saw the cross follow the sword. Conquest was indeed routinely followed by methodical construction of monasteries and churches as part of the process of incorporation of new polities.

Of course, from time to time the Ethiopian elites failed to strike the right balance between expansion and consolidation, between centralization and autonomy. Whenever the delicate balance was questioned the monarchy vacillated.[80] Such was the situation at the time of the confrontation between Ménās and Yeśhāq.[81] Needless to say, the Jesuits contributed to the conflict by attempting to impose themselves as king-makers and promising their sympathizers Portuguese intervention in their favor. Ultimately the Jesuits aimed at projecting the image of a new Catholic Ethiopia in which the heretic nobility and clergy would ultimately lose their prerogatives and material privileges in favor of their Catholic counterparts.[82]

The Ethio-European alliance had reached a breaking point and a common hatred for Islam was no longer sufficient to secure its survival. In fact, so different was the world that Jesuits and Ethiopian elites were remaking that Yeśhāq sought Ottoman support against Ménās in exchange for territories in the vicinity of Massawa.[83] After having been disappointed by the lack of Portuguese support and having seen his allies defeated and killed, Yeśhāq turned to the Ottoman Empire, whose forces waged war against Ménās in 1562. While the Ottomans would soon be defeated, their very intervention allowed the Jesuits to flee from the imperial camp and gain Yeśhāq's protection.[84] The *Bahr Negus* succeeded in offering a safe haven to the remaining Jesuits—who appeared to him as precious mediators with the Portuguese monarchy—but at the same time his policy of collaboration with both Ottomans and Portuguese proved disastrous. The notion of an alliance cutting through religious boundaries was a largely unwelcome novelty both in Portuguese and Ethiopian circles. As a result, Yeśhāq became increasingly isolated and died while attempting to defy Ménās's successor Śarsa Dengel (1563–1597).[85]

The confrontation between Ménās on one side and Yeśhāq with the Portuguese and the Ottomans on the other speaks volumes about the end of an era of Ethiopian and European cooperation. During their sojourn, the Jesuits had managed to negotiate their way into the upper strata of the Ethiopian system and redefine the realm of the possible. In contradiction with the centuries-old discourse on *Prester John* and possibility of a Christian alliance capable of withstanding the Muslim menace, the Portuguese had de facto consented to an indirect alliance—insofar as Yeśhāq negotiated with both powers—with the Ottoman Empire to the detriment of *Prester John*. By the same token the Ethiopian emperor had witnessed the betrayal of fellow Christians. In light of these unexpected turn of events, Rodrigues's 1555 warning that "the notables of the empire would prefer to be subjects to Muslim rule rather than replace their customs with ours"[86] had proven right. Yeśhāq's and Jesuit initiatives had determined the end of an era: both *Prester John* and the *Frank King* of, respectively, European and Ethiopian imagination were now dead. To the contrary, the *Negusa Nagast*, had survived his confrontation with the Muslim foe: Ménās died of natural causes in 1563 and was succeeded by Śarsa Dengel, who successfully weathered through the storm and finally defeated the Ottomans in 1580.

With the patriarch's death—Oviedo had acquired the title after Barreto's death in Goa in 1562—the Jesuit presence in Ethiopia rested solely on the shoulders of António Fernandes and Francisco Lópes, the only surviving members of the Jesuit mission. The two Jesuits put aside their missionary pretensions and limited their activities to the caretaking of the existing Ethiopian and Luso-Ethiopian Catholics. With Francisco Lópes's death in 1597, the Society's first mission to Ethiopia was over and would resume only in 1603 with the arrival of a new party headed by Pedro Páez, the longest-serving and most effective Catholic patriarch to be dispatched to Ethiopia.

Páez reached Fremona in 1603 and remained in Ethiopia until his death in 1622. In 1604 he succeeded in reintroducing the Jesuits at court, befriending the newly appointed and reform-minded Za Dengel (1603–1604), who was eager to understand the workings of European absolutism and implement secular and religious reforms that could strengthen the throne. Páez was successful in his attempt to reassert his presence at court because he remained faithful to Loyola's instructions: he presented himself as a purveyor of progress, limited the conversion efforts to the emperor and his family, and understood the need to approach the mission with a wait and see attitude. In fact, Páez's strategy was initially so circumspect as to advise the emperor himself to postpone any major declaration of reform for fear of attracting the ire of the Ethiopian clergy and the more conservative nobility. The patriarch's concerns were vindicated when Za Dengel prohibited the observance of Saturday by proclamation against his advice: the *abun* accused the emperor of apostasy and supported a rebellion against him that resulted in his defeat and death in October 1604.

By virtue of the distance he had placed between Za Dengel's hasty reforms and himself, Páez survived the turmoil that followed his predecessor's death and rapidly introduced himself at the court of the newly crowned Susenyos (1606–1632).[87] The new emperor developed an interest in Catholicism and in its political potential: as Páez's introduced him to the workings of Iberian absolutism, the emperor started to see how religious reform could limit the power of the Ethiopian Church and its supporters in the regional nobility, and facilitate a coveted centralization of power.[88] Páez's incremental approach came to fruition between 1610 and 1620, when many Ethiopian practices were stigmatized as Judaic corruptions, denounced as heresy and replaced by Catholic ones: finally, in 1621, Páez realized the Jesuit dream by converting Susenyos.

The emperor's conversion, made public through proclamation in 1624, was followed by a fierce repression of Ethiopian traditions and institutions. Ethiopian Churches were closed down and then reconsecrated as Catholic, divorce was made illegal, circumcision forbidden, food prohibitions were challenged, and the estates of the Ethiopian Church were reassigned to the Jesuits.[89] Such radical developments turned a peaceful encounter and the possibility of coexistence for different Christian traditions into violent confrontation. The Ethiopian nobility was split between cooperating with the new emperor's Catholicism or rebelling against it, and the country fell into a civil war.

Susenyos's reforms engendered profound turmoil across the Empire: banditry and rebellion grew rampant as the Ethiopian Church found the support of most of the regional nobility. In 1632, after years of struggle, a disillusioned Susenyos, who had failed to reorganize the empire into a more centralized entity, abdicated in favor of his son Fasiledas (1632–1667) who promptly reestablished Ethiopian orthodoxy. Less than a century after Oviedo's arrival in Massawa, the Jesuits were again banned from court; they were also persecuted and expelled from their stronghold in Fremona. Their Ethiopian enterprise was over: abandoned to itself the Luso-Ethiopian community would survive for a few more years before largely disappearing.[90]

After the expulsion of the last Jesuits from Ethiopia in 1634, the Society was barred from engaging in further missionary activities in Ethiopia and a Papal decree assigned to the newly founded office of Propaganda Fide exclusive jurisdiction over any future activity in the country. While Ethiopian historiography considers the period between the expulsion of the Society and the first wave of modern exploration in the late eighteenth century one of isolation for Ethiopia, Catholic sources relate several attempts to reconnect with the Ethiopian rulers. The expulsion of the Jesuits and the decline of the Portuguese empire did not dissuade the Catholic Church from seeking a way to regain the country. In fact, attempts continued despite several incidents and the death of numerous missionaries who tried to enter Ethiopia between the seventeenth and early eighteenth centuries.[91]

The Jesuit Production of Knowledge

Despite the failure of the mission, there are aspects of the Jesuit experience in Ethiopia that resonated beyond the highlands. Many fathers produced knowledge that redefined the European discourse about Ethiopia and effectively determined the death of *Prester John* in European imagination. The Jesuit period gave birth to the most complete accounts on Ethiopian history and society as well as to a plethora of translations of European and Ethiopian religious texts. It has been argued that through translation the Jesuits sought to export the Counter-Reformation to Ethiopia to facilitate the *reduction*.[92] Those sectors of the Ethiopian elites who supported the Jesuit cause were instrumental in the process of translation by offering assistance. In some cases translations went beyond the realm of the sacred, signaling the Ethiopian perception of the Jesuits as purveyors of progress: Susenyos, for example, demanded the translations of Portuguese laws for the purpose of modernizing the Ethiopian legal system.[93]

Páez edited a bilingual version of the *Cartilha*, a catechism manual that was used to educate children. According to Jesuit correspondence, the text was a great success both in terms of educating the new generations—one of Loyola's original objectives—and impressing the older ones in the process. An Ethiopian nobleman asked the learned children: "How is it that so fast you have been taught so many and good things? Our fathers never teach anything nor make any other good but eating and drinking."[94] We can see *Cartilha* and similar texts as satisfying the double purpose of teaching the tenets of the Catholic tradition and facilitating the Latinization of the country—the abandonment of *ge'ez* as the official language of the Ethiopian Church and the introduction of Latin as the language of the learned elites.

Oviedo compiled a treatise entitled *De Romanae Ecclesiae Primaut, deque errorbus Abassinorum*: while there are no known surviving copies of the volume, we know from references in Jesuit correspondence that Oviedo translated his work in Ge'ez to facilitate consumption on the part of educated Ethiopians.[95] During his sojourn, Méndes compiled *Light of Faith*, a compendium of Ethiopian "errors"[96] that was also translated in Amharic. Father António Fernandes compiled *Life of the Holiest Virgin Mary* and the *Magseph Assetat, Contra Libellum Aethiopicum*, accounts meant to foster the acceptance of Catholic orthodoxy, and he also translated numerous liturgical texts, including masses, prayer books, and catechism.[97]

The Jesuits did not limit their religious writings to liturgical translations and denunciations of heresy; they also translated and reinterpreted hagiographies of Ethiopian saints to justify the Society's presence on Ethiopian soil. Páez, Manuel de Almeida and Barradas translated and commented on a number of lives of saints.[98] Their purpose was to adapt the texts to their missionary efforts, removing references to troubling Ethiopian practices such as circumcision.[99] The Jesuits attempted to demonstrate that Ethiopian Christianity originally conformed to Catholic teachings, and that heresy emerged in Ethiopia only in the eighth century when Egyptian and Semitic thought contaminated the Ethiopian tradition.[100] Furthermore, through the manipulation and reinterpretation of the lives of saints, the Jesuits attempted to demonstrate that in fact the paladins of Ethiopian piety had conducted lives consistent with Catholic teachings. This revision effort can partially be regarded as the natural extension of Loyola's strategy of elite co-optation. Were not Ethiopian saints the ultimate notables of the Ethiopian religious system? If so, the normalization of their hagiographies was a postmortem *reduction* of these paladins of Ethiopian Christianity: by making dead saints consistent with the Catholic canon, the Jesuits hoped to also foster the *reduction* of the living.

Of course, while these hagiographic texts were meant for Ethiopian consumption, the production of Jesuit histories of Ethiopia was instead geared toward the European public. Some of these works—which scholars today correctly regard as hagiographical histories of the Jesuits in Ethiopia[101]—went to occupy a central place in the European Library as demonstrated by the numerous translations and editions offered between the seventeenth and eighteenth centuries. More importantly, they also imposed a new discourse on Ethiopian difference.

The first Jesuit to author a multivolume account on Ethiopia was Páez, with his *Historia de Ethiopia* in 1622. It was by far the most exhaustive collection of historical and contemporary information collected about the country. We know that it served as guide to Páez's successor Méndes and that it was abundantly plagiarized by successive Jesuit accounts. For an array of reasons, most of which are unknown, the work failed to reach the printing press until the early twentieth century. Méndes certainly bears some responsibility, as he judged the work unfit for publication and ordered Manuel de Almeida (1579–1646) to author a new history.[102] The censorial act speaks to the reach of the Society's control over the information produced in the encounter, and confirms its complete grasp over the narrative of the encounter itself—past and present.[103] Other Jesuits compiled more works in the last years of the Jesuit mission: Manoel Barradas (1572–1646) in Ethiopia for more than a decade was responsible for authoring the *Tractatus Tres Historichi Geographici*: like Almeida he completed his work after leaving Ethiopia in 1633. By the same token, the last resident patriarch, Afonso Méndes, authored his *Expeditionis Aethiopicae patriarchae Alphonsi Mendesii* after the expulsion from the country. While these treatises offer great insight into the Jesuit mind, the work that most contributed to a new image of Ethiopia was Almeida's *Historia de Ethiopia Alta ou Abassia*.

Almeida started to work on his *Historia* while in Ethiopia—between 1624 and 1633—but he finished it only in 1644 in Goa following up on Méndes's request. His work became the standard work on Ethiopia as it was the only one among all Jesuit treatises to make it through the Society's censorship. In 1660 *Balthasar Telles* (1596–1675), Jesuit Provincial of Portugal, judged the opera safe for publication and published himself an abridged version in Lisbon. In a matter of years Almeida's book would be made available to the European reading public in a variety of languages and editions. The work was compiled less than a century after Álvares's, however Almeida's discourse was diametrically opposed to that of the Portuguese chaplain.

In his multivolume work Almeida offers a new type of information about Ethiopia: protoethnographical accounts of what he referred to as different "races" and comments on Christian Ethiopians in the following terms: "the Abyssinian generally have well-shaped figures, good height and good facial features, spare bodies, pointed noses and thin lips, so that the people of Europe have the *advantage* of them in colour but not in other things."[104] Despite the racialization of the discourse, Almeida's characterization of Ethiopians is still far from racial determinism, as he quickly adds that "[Ethiopians] are not cruel or bloodthirsty."[105] Nevertheless, when it comes to the construction of otherness, the die is cast: Ethiopians have turned from Christian brothers into heretics whose physical features denote a certain degree of inferiority and at least a modicum of undesirability. Furthermore, with regard to baptismal practices and circumcisions, Almeida is even more vocal about Ethiopia's remoteness from the Catholic norm and comments that

> It will be apparent from this how many souls have lost Heaven through this error in the course of so many hundreds of years. Today, after they have received the holy Roman faith, one of the things they cannot be

> persuaded to do is to abandon circumcision. They say they do not do it to keep the law of Moses but only for elegance. Great folly or blindness![106]

Similarly condescending is Almeida's characterization of Ethiopian marriages and in particular his criticism of the Ethiopian practice of *damoz*, a temporary form of domestic union:

> Until our own time weddings among the Abyssinians were of such a kind that there was never true marriage between them, because they married with the tacit or expressed agreement to dissolve it as soon as the husband and wife disagreed with one another. For this reason they used to exchange guarantors with certain peculiar and barbarous customs, for these people, who were together like paranymphs, had to be almost eyewitnesses of the consummation of the marriage.[107]

Some attempted to explain the stark difference between Álvares and Almeida's accounts by advancing the claim that Álvares's peculiar relativism was simply the result of his ignorance and lack of preparation. According to this argument Álvares was unprepared to properly assess life in remote Ethiopia and lacked the cultural and theological knowledge necessary to pass proper judgment.[108] The blissful ignorance of the first Portuguese encounter, the argument continues, would later be zealously overcome through Jesuit competent intervention.

The argument does not wash for a simple reason: Álvares's account was not the product of Europe's first confrontation with Ethiopian traditions. Ethiopians had been described as followers of peculiar religious practices for more than two centuries. While it is correct to argue that knowledge production was scarce until Álvares's account and that it spiked only decades later under Jesuit leadership, the pre-Jesuit period of the Luso-Ethiopian encounter offered more than enough exotic material to warrant a formulation of Ethiopian otherness. Instead, it is as peers that Ethiopians were described both in *Verdadera* and in related writings. Pre-Jesuit authors refrained from attributing any moral meaning to the practices and customs they witnessed. Hence, Álvares's objective and detached descriptions of the Ethiopian way of life, far from being the result of ignorance on his part, should instead be regarded as the ultimate confirmation of a commendably relativist attitude toward Ethiopia before the birth of a Eurocentric discourse under Jesuit midwifery. To the contrary, the sampled evidence from Almeida's *Historia* is a very telling example of the shift toward marked Eurocentrism. One can argue that this new attitude toward difference was simply the result of protracted exposure and increased chances of observation: according to this argument a more stable cultural encounter on Ethiopian soil meant a more refined understanding of each other and more occasions for European scrutiny of Ethiopian culture. However, evidence suggests that rather than representing the Jesuits' overcoming of blissful ignorance, the new discourse of difference and the replacement of Ethiopian Christian sameness with heretic otherness was the brainchild of militant proselytism.

Conclusion

While hinged on understandings of religious identity—for which true faith needed to be defended against religious heresy—the Jesuit discourse on Ethiopia appears a far cry from the same discourse in the pre-Jesuit era. The Society's agents replaced the Muslim-Christian divide with the Catholic-heretic one and, in the process, they also created Ethiopian *otherness*: the *Prester* and his countrymen's religious beliefs and practices were to be normalized by peaceful means when possible and through violence if necessary. It has been argued that the Jesuits were caught between *accomodatio* and *impositio*, between the word and the sword:[109] our brief survey suggests that on Ethiopian soil imposition accounted for the lion share of the Society's missionary efforts and that the latter would have turned even more violent had the Portuguese authorities been more supportive of the effort. During their protracted attempt to export Catholic salvation to Ethiopia, many Jesuits lobbied the Portuguese crown and the upper echelons of the Estado de India to offer military support and deploy soldiers to facilitate the process of conversion.

The contemplation of a military option, the production of knowledge both in the form of pro-*reduction* theological propaganda and historical narrative, the co-optation of local elites according to a divide et impera script, the desire to educate the youth and uproot it from tradition, all seem to point to the adoption of strategies that would later become the staple policies of nineteenth-century empire-building. It is perhaps a simplistic exaggeration to dub the Society the first multinational corporation in an effort to underline its modernity,[110] but the Jesuits in Ethiopia certainly demonstrated to be well ahead of their time by recognizing that, in dealing with *others*, violence—the sine qua non of modern imperialism—was a necessary evil. The Jesuits in Goa never convinced the Portuguese Crown to draw the sword against the Ethiopian Empire, yet by the early seventeenth century the *Prester* of the early modern European imagination had already fallen to the Jesuit pen.

Notes

Portions of this chapter have been presented at the 2009 World History Association Conference in Salem, MA. I would like to express my gratitude to David Northup and the two anonymous reviewers for their valuable comments. The mistakes are solely mine.

1. Loyola to Galawdéwos (21-02-1555) in Maurice Giuliani, *Ecrits* (Paris: Desclee de Brouwer, 1991), 918–922.
2. For a detailed overview of the structure and history of the Portuguese *assistancy* see Dauril Alden, *The Making of an Enterprise: The Society of Jesus in Portugal, Its Empire, and Beyond, 1540–1750* (Stanford: Stanford University Press, 1996). More in general on the structure and workings of the Jesuit orders the most recent work with a true global perspective is Luke Clossey, *Salvation and Globalization in the Early Jesuit Missions* (New York: Cambridge University Press, 2008).
3. In the Ethiopian context the Society of Jesus intended the process of reduction as the abandonment, on part of Ethiopians, of heretic practices and theological beliefs and the return to the true Catholic faith.
4. For a discussion of the first Ethiopian travelers in Renaissance Europe and the emergence of an early modern image of Ethiopia as the Christian land of Prester John see Matteo Salvadore, "The Ethiopian Age of Exploration: Prester John's Discovery of Europe, 1306–1458." *Journal of World History* 21, no. 4 (2010): forthcoming.
5. For a thorough discussion of Ethiopian presence in Europe in the fourteenth century see P.M. Chaine, "Un monastère éthiopien a Rome au XVe et XVI siècle," *Mélanges de la Faculté orientale* 5 (1911): 1–36; Mauro Da Leonessa, *Santo Stefano Maggiore degli Abissini e le relazioni romano-etiopiche* (Rome: Vatican, 1928), Renato Lefevre, "Riflessi etiopici nell cultura europea del medioevo e del rinascimento," *Annali Lateranensi* 11 (1947): 255–342. For the Ethiopian presence in Jerusalem see Enrico Cerulli, *Etiopi in Palestina Storia Della Comunita' Etiopica Di Gerusalemme* (Roma: Libreria dello Stato, 1943).
6. Dates in parenthesis indicate birth and death with the exception of institutional figures for which dates indicate time in office.
7. Camillo Beccari, *Rerum aethiopicarum scriptores occidentales inediti a saeculo XVI ad XIX*, vol. 5 (Romae: C. de Luigi, 1903), 357–358.
8. On the figure of Andre' de Oviedo see Eduardo Janvier Alonso Romo, "Andrés de Oviedo, patriarca de Etiopía," *Penisula, Revista de Estudos Ibéricos* no. 3 (2006): 215–231.
9. Lefevre, Documenti Renato Lefevre, "Documenti pontifici sui rapporti con l'Etiopia nei secoli XV e XVI," *Rassegna di Studi Etiopici* 5 (1947): forthcoming.
10. *Färäng* is an Amharic term borrowed from Arabic in the Middle Ages. At first it identified Catholics and later it came to identify any person of European descent. See "Färäng" in *Encyclopaedia Aethiopica Encyclopaedia Aethiopica*, vol. 1 (Wiesbaden: Harrassowitz Verlag, 2003), 492.
11. For a discussion of early modern Ethio-European contacts in the context of Afro-European relations see Werner Debrunner, *Presence and Prestige, Africans in Europe: A History of Africans in Europe before 1918* (Basel: Basler Afrika Bibliographien, 1979); David Northrup, *Africa's Discovery of Europe: 1450–1850* (New York: Oxford University Press, 2002); Adrian Hastings, *The Church in Africa: 1450–1950, The Oxford History of the Christian Church* (Oxford: Oxford University Press, 1994). About the early fifteenth

century's Ethiopian missions to Venice see Kate Lowe, "'Representing' Africa: Ambassadors and Princes from Christian Africa to Renaissance Italy and Portugal, 1402–1608," *Transactions of the Royal Historical Society* 17 (2007): 101–128; Marilyn E. Heldman, "A Chalice from Venice for Emperor Dawit of Ethiopia," *Bulletin of the School of Oriental and African Studies* 53, no. 3 (1990): 442–445.

12. The letter was first mentioned in the Chronicle of Otto of Freising: Otto von Freising. *Chronica Sive Historia De Duabus Civitatibus* (Germany: 1143), 266.
13. Among them are fragments from the accounts of Pliny, St. Augustine, Isidore of Seville as well as the apocryphal *Romance of Alexander.* Malcolm Letts, "Prester John: A Fourteenth-Century," *Transactions of the Royal Historical Society* 29 (1947): 20; Marshall Monroe Kirkman, August Petrtyl, Cropley Phillips Company, and R.R. Donnelley, *The Romance of Alexander the King: Being One of the Alexandrian Romances, "Alexander the Prince", "Alexander the King" & "Alexander and Roxana"* (London: Cropley Phillips, 1909). Zarncke, *Der Priester Johannes,* vol. 7 (1879), 941. See also Charles M. de La Roncière, *L'Europe au Moyen âge* (Paris: A. Colin, 1969), 57.
14. The so-called Third Crusade, which targeted the Egyptian Sultanate, was crashed by Salah ad-Din Yusuf ibn Ayyub (1137–1193). See John Larner, *Marco Polo and the Discovery of the World* (New Haven: Yale University Press, 1999), 14.
15. Dispatched respectively by Pope Innocent IV (1243–1254) and Louis IX (1226–1270) of France.
16. J.L. Abu-Lughod, "The World System in the Thirteenth Century: Dead-End or Precursor?" in *Islamic & European Expansion: The Forging of a Global Order* (Washington, DC: American Historical Association, 1993), 76.
17. I borrow the concept from Mary Louise Pratt, *Imperial Eyes: Travel Writing and Transculturation* (New York: Routledge, 1992), 9–10.
18. N. Jorga, "Cenni sulle relazioni tra l'Abissinia e l'Europa cattolica nei secoli XIV–XV, con un itinerario inedito del secolo XV," in *Centenario della nascita di Michele Amari,* 2 vols. (Palermo: 1910), 1:142.
19. See Renato Lefevre, "Riflessi etiopici nell cultura europea del medioevo e del rinascimento," *Annali Lateranensi* 11 (1947): 255–342
20. Arch. Societas Iesus, Goana Hist. Aeth. 1549–1629, Doc. IV ff8–12, in Beccari, *Rerum aethiopicarum,* 1:237.
21. Niccolò da Poggibonsi, *Libro d'oltremare, scelta di curiosità letterarie inedite o rare dal secolo XIII al XVII* (Bologna: G. Romagnoli, 1881).
22. Interest in Prester John was already a factor at the time of Prince Henry of Portugal's feats in northwestern Africa and later became one of Afonso V's (1438–1481) priorities. For an overview see M.D.D. Newitt, *A History of Portuguese Overseas Expansion, 1400–1668* (London: Routledge, 2005) and P.E. Russell, *Prince Henry "the Navigator:" A Life* (New Haven: Yale University Press, 2000).

23. For an overview of Portuguese exploration and settlement on the African coasts see Russell, *Prince Henry,* 125–130; Newitt, *A History,* 27 and the original chronicle by Gomes Eanes de Zurara, C. Raymond Beazley, and Edgar Prestage, *The Chronicle of the Discovery and Conquest of Guinea* (New York: B. Franklin, 1963).
24. Matéwos was either an Armenian or Egyptian ambassador at Eleni's service. The account that resulted from Góis's editing of Matéwos's confession of faith together with the official letters exchanged between the Portuguese and Ethopian sovereigns was first published in 1531 in Antwerp under the title *Legatio Magni Indorvm Imperatoris Presbyteri Ioannis, ad Emanuelem Lusitaniae Regem, Anno Domini MDXIII*: Góis was unaware of its publication, which seemed to have been the initiative of Johannes Mansson (1488–1544), the exiled Swedish archbishop Góis corresponded with. Góis's collaboration with Zagā Za'āb resulted instead in the publication of Damião de Góis, ed. *Fides religio moresque Aethiopum sub imperio Preciosi Ioannis dentium* (Louvain: Rescius, 1540). The account was also published under the title *The faith, religion and manners of the Aethiopians,* as appendix to Joannes Boemus, Nicolaus, Jean de Léry, Edward Aston, George Eld, and English Printing Collection (Library of Congress), *The manners, lauues, and customes of all nations* (London: Printed by George Eld, 1611).
25. Andrea Corsali, *Lettera di Andrea Corsali all'Ill.mo Signore Duca Juliano de Medici venuta dell'India nel mese di octobre nel MDXVI and Lettera di Andrea Corsali allo Ill.mo principe et signore Laurentio de Medici duca d'Urbino ex India* (1516).
26. Diogo Lopes de Sequeria, Pero Gomes Teixeira, *Carta das novas que vieram a el rei mossa senhoe do descobrimento do preste Joham* (Lisboa: 1521, 1938).
27. The text was prepared for publication during the years between Álvares's return to Lisbon in 1533 and his death in 1540. For a discussion of Álvares's whereabouts see the introduction to his narrative in Francisco Álvares, and Henry Edward John Stanley Stanley, *Narrative of the Portuguese Embassy to Abyssinia, during the Years 1520–1527* (New York: B. Franklin, 1964), 7–9. The chaplain's plan seems to have been a much longer country study in five volumes with sections on its geography, soil and natural resources, animals, customs, institutions and last but certainly not least, faith. His plan never came to fruition and instead he published an impressive yet haphazardly organized account that seem often at odds with each other. Most likely the chaplain hoped to publish a more structured and detailed account in later years, possibly by integrating his notes with information he expected to obtain from the Ethiopian ambassador he escorted to Lisbon, Zagā Za'āb. Álvares's sudden death in 1540 shuttered his plans and Zagā Za'āb collaborated instead with Góis.
28. The first English translation was completed by Baron Stanley of Alderly in 1881 for the Hakluyt Society. A Spanish version was published first in Antwerp in 1557 and later in Toledo in 1588. Ramusio included an Italian version in his *Navigationi et Viaggi,* published in Venice in 1550 and an Italian cleric, Ludovico Beccadelli included portions of it in a 1526. For

a discussion of the different versions see Francisco Alvares, *Narrative of the Portuguese Embassy to Abyssinia,* 5-6 and Roberto Almagià, *Contributi alla storia della conoscenza dell'Etiopia* (Padova: University of Padova Press, 1941), 14–17. For a discussion of Hakluyt and Purchas's contributions to the development of British Imperial ideology see David Armitage, *The Ideological Origins of the British Empire* (Cambridge and New York: Cambridge University Press, 2000).

29. The version used for the analysis is the latest English translation published by the Hakluyt Society. Alvares, *Narrative,* 38.
30. Ibid., 286.
31. Ibid., 288.
32. On the Ethiopian-Adali confrontation see the Adali chronicle and the account of Miguel de Castanhoso, officer in the Portuguese battalion. Sihab ad-Din Ahmad bin Abd al-Qader bin Salem bin Utman, *Futuh Al-Habasha: The Conquest of Abyssinia* (Tsehai, 2003); R.S. Whiteway et al., *The Portuguese Expedition to Abyssinia in 1541–1543 as Narrated by Castanhoso, with Some Contemporary Letters, the Short Account of Bermudez, and Certain Extracts from Corrêa* (Nendeln, Liechtenstein: Kraus Reprint, 1967).
33. See Mauro Da Leonessa, *S. Stefano dei Maggiore o degli Abissini* (Rome: Vatican, 1928).
34. Zagā Za'āb's manuscript was originally published by Damião de Góis as *Fides religio moresque Aethiopum sub imperio Preciosi Ioannis dentium* in 1540 and later in appendix to Joannes Boemus et al., *The manners, lauues, and customes of all nations* (London: Printed by George Eld, 1611); Damião de Góis, ed. *Fides religio moresque Aethiopum sub imperio Preciosi Ioannis dentium* (Louvain: Rescius, 1540).
35. For a profile of Góis and his involvement with Ethiopia see Elizabeth Feist Hirsch, *Damião de Góis: The Life and Thought of a Portuguese Humanist, 1502–1574* (Antwerp: Martinus Nijhoff, 1967): Jeremy Lawrance, "The Middle Indies: Damião de Góis on Prester John and the Ethiopians," *Renaissance Studies* 6, nos. 3–4 (1992): 316–324; Jean Aubin, *Le latin et l'astrolabe. Recherches sur le Portugal de la Renaissance, son expansion en Asie et les relations internationales* (Paris: Fundaçao Calouste Gulbenkian, 1996).
36. Góis authored *Legatio Magni Indorvm Imperatoris Presbyteri Ioannis, ad Emanuelem Lusitaniae Regem, Anno Domini MDXIII* and included numerous pages about Ethiopia in his Damião de Góis, *Crónica do felicíssimo rei D. Manuel* (Lisboa: 1566).
37. Hervé Pennec, *Des jésuites au royaume du prêtre Jean, Ethiopie: stratégies, rencontres et tentatives d'implantation, 1495–1633* (Paris: Centre culturel Calouste Gulbenkian, 2003), 24. See also Andreu Martínez Alòs-Moner, "The Birth of a Mission: The Jesuit Patriarchate in Ethiopia," *Portuguese Studies Review* 10, no. 2 (2003): 1–14; and Andreu Martinez D'Alos Moner, "Paul and the other," in Verena Böll et al., *Ethiopia and the Missions: Historical and Anthropological Insights* (Münster: Lit, 2005), 35–36.
38. Elisabeth Feist Hirsch, *Damião de Góis; the Life and Thought of a Portuguese Humanist, 1502–1574* (The Hague: M. Nijhoff, 1967), 96.

39. Ibid., 190–220.
40. Cited in Leonessa, *S. Stefano*, 253.
41. Leonessa also cites a later episode of an Ethiopian monk—Gebre Krestos—who wrote to Propaganda Fide to report the detail of his ordination by the hands of the Egyptian patriarch in Cairo. No other source seems to be available on the fate of the Ethiopian monk, yet the letter shows that the Catholic authorities in Rome had grown vigilant with regard to S. Stefano's guests. Leonessa, *S. Stefano*, 251.
42. About the decreasing number of Ethiopians at Santo Stefano see Leonessa, *S. Stefano*, 252.
43. See De Lorenzi's account of Gorgoryos's arrival in Rome as a Catholic refugee, also discussed in Leonessa, *S. Stefano*, 270–280.
44. Beccari, *Rerum aethiopicarum*, 10:7. The passage is extensively commented in Aubin, *Le latin et l'astrolabe*, 201–203.
45. Renato Lefevre, "Documenti e notizie su Tasfa Seyon e la sua attività romana nel sec. XVI." (1969): 75.
46. Beccari, *Rerum aethiopicarum*, 10:237.
47. Ibid., 1:237.
48. See Loyola's letter in Pennec, *Des jésuites*, 11.
49. For the concept of *voluntary dependence* and its relevance for the stability of the Ethiopian feudal system and its relation with the Egyptian Sultanate see Hagai Erlikh, *The Cross and the River Ethiopia, Egypt, and the Nile* (Boulder: Lynne Rienner, 2002), 20.
50. Da Urbina argued that the letter represented an offer of union and that later the offer was betrayed. In this perspective the offer is presented as one of convenience. I. Ortiz Da Urbina, "L'Etiopia e la Santa Sede nel secolo XVI," *Civilta' Cattolica* 4 (1934): 388. Lefevre also argues that the letters represent a declaration of obedience on part of the Ethiopian Emperor. Renato Lefevre, *L'Etiopia nella stampa del primo Cinquecento* (Como: Cairoli, Edizioni africane, 1966), 54. Álvares had acted as the translator and editor of the letters that reached Clement VII, a circumstance that suggest further uncertainty with regard to the original wording and intention of the Ethiopian elites. Furthermore we know from Zagā Za'āb's experience in Lisbon and his declaration of faith that the majority of Ethiopian elites where unwilling to submit and instead saw Rome as a primus inter pares.
51. Beccari, *Rerum aethiopicarum*, 1:238.
52. There are several accounts of Bermudes's feats, among which his own account, published in R.S. Whiteway, Miguel de Castanhoso, João Bermudez, and Gaspar Corrêa, *The Portuguese Expedition to Abyssinia in 1541–1543 as Narrated by Castanhoso, with Some Contemporary Letters, the Short Account of Bermudez, and Certain Extracts from Corrêa* (Nendeln: Kraus Reprint, 1967).
53. See Loyola's letter in Beccari, *Rerum aethiopicarum*, 5:55–56.
54. Cited in Andreu Martínez Alòs-Moner, "The Birth of a Mission," 10. The Bull's text can be found in Beccari, *Rerum aethiopicarum*, 10:39–41.

55. Ibid., 11.
56. According to Lebna Dengel's letter to Manuel I "[the Portuguese embassy to Ethiopia] was first prophesied by the prophet in the life and passion of St Victor, in the book of the Holy Fathers [...] that a Frank King should meet with the King of Ethiopia, and that they should give each other peace," see Alvarez, *The Prester John,* 60. In the same page we can find Álvares's own interpretation of the prophecy: "The Abyssinians had a prophecy that there would both be more than a hundred Popes in their country, and that then there would be a new ruler of the Roman Church and that the Abima would complete the hundred; and also they had two prophecies one of St Ficatorio, the other of St Sinoda who was a hermit of Egypt, saying that the Franks from the end of the earth would come by sea and would join with the Abyssinians and would destroy Juda [Jedda], and Tero [Tor] and Meca [Mecca] and that so many people would cross over and would pull down Meca, and without moving would hand the stones from one to another and would throw them into the Red Sea, and Meca would be left a bare plain, and that also they would take Egypt and the great city of Cairo," Ibid.
57. Beccari, *Rerum aethiopicarum,*1:240.
58. Since the time of Dawit's request of artisans in Venice, to continue with Lebna Dengel's letter to the Pope containing a request for skilled workers, all the way to the nineteenth-century Ethiopian emperors' penchant for European weaponry in the late nineteenth century, technological transfer remained a constant Ethiopian interest throughout the modern era.
59. Beccari, *Rerum aethiopicarum,* 1:239 and 250.
60. Merid W. Aregay, "Two Inedited Letters of Galawdéwos," *Studia* (1964): 13–14 and 372–373.
61. Beccari, *Rerum aethiopicarum,* 10:78–81.
62. Quoted in Pennec, *Des jésuites,* 99.
63. See two letters in Beccari, *Rerum aethiopicarum,* 10:215–220 and also Beccari, *Rerum aethiopicarum,* 3:71–75 and Beccari, *Rerum aethiopicarum,* 5:427–432.
64. Beccari, *Rerum aethiopicarum,* 10:332–334.
65. Pero Páez has been featured prominently in both Philip Caraman, *The Lost Empire: The Story of the Jesuits in Ethiopia, 1555–1634* (Notre Dame: University of Notre Dame Press, 1985), and Pennec, *Des jésuites,* which also offers the organizational details of the Jesuit presence in Ethiopia. Caraman is overtly hagiographical as he portrays Páez in saintly and innocent terms while scapegoating Mendez for the implementation of counterproductive policies that lead to the Jesuit expulsion. For a scathing critique of Páez's tenure as Catholic patriarch see Merid Wolde Aregay, "The Legacy of Jesuit Missionary Activities in Ethiopia," in Getatchew Haile, Samuel Rubenson, and Aasulv Lande, eds. *The Missionary Factor in Ethiopia: Papers from a Symposium on the Impact of European Missions on Ethiopian Society* (Frankfurt am Main: Verlag, 1998), 31–56. Merid Wolde Aregay is the most vocal apologist of Mendez and presents Páez in a less favorable light than previous historiography. For another sympathetic account of

the Jesuit mission see Mauro Da Leonessa, *Santo Stefano Maggiore degli Abissini e le relazioni romano-etiopiche* (Rome: Vatican, 1928).

66. I am borrowing the concept from Pennec, *Des jésuites,* Chapter 3.
67. F.M. Esteves Pereira, *Chronica de Susenyos, rei de Ethiopia* (Lisboa,: Imprensa nacional, 1892), 259.
68. See the discussion of the issue of lime-masonry as novelty in Pennec, *Des jésuites,* 175–180.
69. Augusto Salimbeni, "Tre anni di lavoro nel Goggiam," *Bollettino della Società geografica italiana* 2, VXI (1886): 279–280.
70. For an overview of province-specific strategies see Alden, *The Making.*
71. See Loyola's *Instructions* in Beccari, *Rerum aethiopicarum,* 1:242.
72. See Loyola's *Instructions* in ibid., 1:251.
73. Ibid.
74. Almeida in Beccari, *Rerum aethiopicarum,* 5:383–384.
75. For a discussion of Ethio-Egyptian relations see Mordechai Abir, *Ethiopia and the Red Sea: The Rise and Decline of the Solomonic Dynasty and Muslim-European Rivalry in the Region* (London: Totowa, NJ: F. Cass, 1980) and Hagai Erlikh, *The Cross and the River Ethiopia, Egypt, and the Nile* (Boulder, CO: Lynne Rienner, 2002).
76. Girma Beshah and Merid Wolde Aregay, *The Question of the Union of the Churches in Luso-Ethiopian Relations, 1500–1632* (Lisbon: Centro de Estudos Históricos Ultramarinos, 1964), 61–63.
77. *Bahr Negus,* literary "King of the Sea," is the traditional title of a regional Ethiopian ruler whose jurisdiction extended on coastal Ethiopia and Tigray. See "Bahär Nägaš" in *Encyclopaedia Aethiopica,* vol. 3 (Wiesbaden: Harrassowitz Verlag, 2003), 444.
78. The size of the Catholic community has been estimated at about one thousand people. See Pennec's estimates in Pennec, *Des jésuites,* 104.
79. For a detailed discussion of the "Church and State" complex see Taddesse Tamrat, *Church and state in Ethiopia, 1270–1527* (Oxford: Clarendon Press, 1972).
80. In the early modern period there were two examples of crisis for the Church and State complex: the difficult quest for a new *Abun* under Be'eda Maryam (1468–1478) and the Adali invasion mentioned earlier in the chapter. In the first case the lack of leadership in the Church led to difficult dynastic successions; in the second turmoil along the border with Adal had been largely caused by a rapid and disorganized expansion in the fifteenth century. See Taddesse Tamrat, *Church and State,* 198–200.
81. For a detailed discussion of Ethiopian politics in the Jesuit era see Beshah and Aregay, *The Question* and Taddesse Tamrat, "Evangelizing the Evangelized: The Root Problem between Missions and the Ethiopian Orthodox Church," in Haile, Rubenson, and Lande, *The Missionary Factor in Ethiopia.*
82. Contrapositions between supporters of ecclesiastical independence and unifications had characterized most of the fifteenth century. See Aubin, *Le latin et l'astrolabe,* 141–150; Tamrat, *Church and State in Ethiopia*; Pennec, *Des jésuites,* 29–30.

83. Merid Wolde Aregay "The Legacy of Jesuit Missionary Activities in Ethiopia," in Haile, Lande, and Rubenson, *The Missionary Factor in Ethopia*, 42.
84. Aregay "The Legacy," 64.
85. Ibid., 68.
86. Beccari, *Rerum Aethiopicarum*, 5:357.
87. A period of factionalist confrontation followed Za Dengel's death. Before the emergence of Susenyos, Yacob briefly regained the throne between 1604 and 1605, but his imperial tenure was brief and ephemeral.
88. The argument is developed in Aregay, *The Legacy*, 43–48.
89. Beshah and Aregay, *The Question*, 92–93.
90. Until 1648 when Abun Marqos organizes a punitive expedition that resulted in the death and displacement of all Catholic of Ethiopian and mixed descent from Tigray. Beccari, *Rerum aethiopicarum*, 13:322 and 380.
91. Three fathers were executed in Suakim by order of the Ethiopian Emperor in 1662. Antonio d'Andrade, Vicar Apostolic, and two Franciscans were killed in Massawa in 1671. In 1797, Ignazio Ballerini, a priest, was killed in Nubia while trying to enter Ethiopia. For an overview of the postexpulsion attempts to reenter Ethiopia and the related deaths see Beccari, *Rerum aethiopicarum*, 1:175–185.
92. The most indicative example of this policy is the translation of Cardinal Bellarmine (1542–1621) anti-Protestant writings. Leonardo Cohen, "The Jesuit Missionary as a Translator (1603–1632)," in Böll, *Ethiopia and the Missions*, 8.
93. Cohen, "The Jesuit Missionary," 10.
94. Ibid., 16. The letter is available in Beccari, *Rerum aethiopicarum*, 3:237–238.
95. Eduardo Janvier Alonso Romo, "Andrés de Oviedo," 226; See Beccari, *Rerum aethiopicarum*, 5:383–384.
96. Afonso Mendes, *Bran Haimanot: Id est Lux Fidei in Ephitalamium Aethiopissae sive in Nuptias Verbi et Ecclesiae* (Lisboa: 1642).
97. Cohen, "The Jesuit Missionary," 12. Pennec, *Des jésuites*, 265–267.
98. Ibid., 18.
99. Ibid., 20.
100. Manuel Barradas, *Tractatus tres historico-geographici (1634): A Seventeenth Century Historical and Geographical Account of Tigray, Ethiopia* (Wiesbaden: Harrassowitz, 1996), Chapter 13 and Chapter 47; Afonso Mendes, and Camillo Beccari, *Expeditionis aethiopicae* (Rome: 1908), Chapter 26.
101. See Pennec's extensive discussion of these sources. Pennec, *Des jésuites*, Chapter 5.
102. For a discussion of the reasons behind the censorship see Pennec, *Des jésuites*, 258–267.
103. Beccari, *Rerum aethiopicarum*, 1:3.

104. Emphasis added. C.F. Huntingford, Manuel de Almeida et al., *Some records of Ethiopia, 1593–1646, being extracts from the History of High Ethiopia or Abassia, by Manoel de Almeida, together with Bahrey's History of the Galla* (London: Hakluyt Society, 1954), 60.
105. Ibid.
106. Ibid., 62.
107. Ibid., 64.
108. This is Beckingham's argument; see his introduction to Francisco Alvares et al., *The Prester John of the Indies* (Cambridge: Cambridge University Press, 1961).
109. Andreu Martínez Alòs-Moner, "Paul and the Other: The Portuguese Debate on the Circumcision of the Ethiopians," in Böll, *Ethiopia and the Missions*, 32.
110. The analogy was first proposed in C.R. Boxer, *Portuguese India in the Mid-seventeenth Century* (Delhi: Oxford University Press, 1980) and discussed in appendix to Dauril Alden, *The Making of an Enterprise.* Boxer argued that the Society was more international and better organized than the Dutch and English India Companies, often regarded as the first example of multinational corporation. Alden's criticism of this analogy points out that economic profit was not the main driver of the Society's activities—which was instead interested in the goal of salvation of souls—and that the decision-making process of the Society privileged diffusion rather than centralization.

CHAPTER 8

Red Sea Travelers in Mediterranean Lands: Ethiopian Scholars and Early Modern Orientalism, ca. 1500–1668

James De Lorenzi

In 1534, an Ethiopian scholar named Ṣagā Za'āb (n.d.–1539) wrote a defense of his Christian faith for the Dutch humanist Damião de Góis (1502–1574). Ṣagā Za'āb had come to Portugal to support the diplomatic mission of his patron, the Ethiopian emperor Lebna Dengel (1508–1540), but upon arrival he found his beliefs challenged by the Catholic theologians of Lisbon. He decided to refute these hostile critics through a text that would document the sound foundations of his faith and correct the Europeans' many misperceptions of his country and people. He began this work with an outline of the main tenets of Ethiopian Orthodox Christianity, supported by scriptural and patristic proofs, and he continued with a critique of the Catholic position and a summary of several texts from his scholarly canon—including the epic *Kebra negest* (Glory of Kings), which purported to document the Hebraic descent of Ethiopia's Christian rulers. For the first time, this African narrative of the Queen of Sheba was related to European readers. After completing his treatise with a plea for true ecumenical fraternity, Ṣagā Za'āb sent his Ge'ez language text to De Góis in Louvain, where the latter published a Latin translation in 1540.[1] Though it was censored within a year, Ṣagā Za'āb's heretical but still immensely informative text circulated among European readers for the next two centuries, and it eventually became a canonical source for early modern scholars of Africa, Asia, and the wider world.

Though this story of cross-cultural intellectual exchange might seem exceptional, it was not. Between the sixteenth and mid-seventeenth centuries, several highly educated Ethiopians visited the urban centers of Mediterranean Europe and introduced their hosts to a previously unknown

African intellectual tradition. These diasporic Ethiopian scholars took up a wide range of projects. Like Ṣagā Za'āb, some focused on the production of printed books: for example, in 1548, an Ethiopian resident of Rome named Tasfā Ṣeyon (n.d.–1550) organized the publication of *Testamentum Novum,* the first printed version of the Gospels in the Ge'ez language.[2] Other Ethiopian visitors acted as information brokers for the growing number of European scholars interested in Eastern Christianity. In 1513, a monk named Tomās Walda Samu'él (n.d.) assisted with the Ge'ez text of a polyglot Psalter, and a century later, a scholar named Gorgoryos (n.d.–1658) collaborated with the German Hiob Ludolf (1624–1704) on a monumental study of the languages and history of Northeast Africa.[3] And some Ethiopian visitors served as advisors to prominent figures in the Catholic Church: Giovanni Battista Abissino (n.d.) counseled Pio IV (1499–1565), and Tasfā Ṣeyon, editor of *Testamentum Novum,* met Ignatius Loyola (1491–1556) on multiple occasions.[4] Just as early modern imperial and missionary networks began to broker a massive influx of information to Europe from Asia and the Americas, and at a moment of intense European interest in the exotic texts and traditions of Eastern Christianity, these Ethiopian scholars began a transmission of specialized knowledge to Europe that profoundly shaped the emerging global imaginary.

Though early modern historians have shown great interest in the dynamics and consequences of cross-cultural encounters, few have noted the activities of the community of Ethiopian scholars in Europe.[5] An earlier generation of Italian specialists meticulously compiled and published archival materials related to Santo Stefano degli Abissini, the Ethiopian diasporic residence in Rome,[6] but several recent studies of European encounters with Africa and the Middle East have not addressed the topic, despite its relevance to our understanding of intellectual and cultural exchange in the early modern period.[7] And the story of Ge'ez language printing in Europe is similarly unknown, though it represents one of the earliest chapters of African and Asian book history.[8] The present chapter engages these literatures by exploring this early modern flow of knowledge from Northeast Africa to Europe, tracing it from origin to final destination. It begins by mapping the intellectual culture of the Red Sea region and outlining its real and imagined links to the wider Mediterranean world. It next explores the role of Ethiopian scholars as transmitters and producers of knowledge in early modern Europe, focusing both on their writings and the social and institutional bases that sustained their work. It concludes by examining the influence of this scholarship on readers, in an attempt to situate these Ethiopian texts in their European and global contexts.

Ultimately, our seemingly obscure story has a number of larger implications. On a basic level, it adds to our understanding of the heterogenous set of actors and institutions involved in European intellectual constructions of the wider world. But more significantly, it illustrates how the changing dynamics of early modern globalization brokered new and connected forms of world-building in Europe, Africa, and Asia: though it is tempting to see this case of cross-cultural exchange as anomalous, it is more fruitful to see it as one of many examples of intellectual borrowing and refusal that Fernand Braudel described as "hallmarks of civilization."[9] Here, exchange produced two different outcomes, each with lasting significance. For Europeans, the intellectual borrowing from Ethiopia laid the foundations for the emergence of the philological, orientalist approach to the study of Africa and Asia. But for Ethiopian Christians, this encounter ultimately led to political rejection and intellectual refusal.

Linked Worlds: The Red Sea and the Early Modern Mediterranean

Ṣagā Za'āb and his learned colleagues emerged from the tumultuous world of sixteenth- and seventeenth-century Northeast Africa. The region was then dominated by conflicts between the Christian empire of highland Ethiopia and the Muslim Adal Sultanate of present-day Somalia and eastern Ethiopia, and both parties enlisted foreign allies in their competing bids for dominance. There was a profound religious dimension to this struggle. Though the power of Ethiopia's Christian emperors (*negusa nagast*) depended upon the arms of the nobility (*čawā*) and the tribute of small landholders (*balagar*), it also relied upon the church itself (*bét kahānat*), whose erudite scholars spread the faith, administered justice, and developed a potent Solomonic ideology that legitimated royal claims to represent and defend Christianity. In turn, the Muslim leaders of the Adal Sultanate, most notably Ahmad b. Ibrahim al-Ghazi (ca. 1506–1543), cast their struggle against the Christian empire as a jihad, thereby attempting to rally Muslims throughout the Red Sea arena. In the 1540s, this conflict led the Portuguese and the Ottomans to military interventions in support of their local coreligionists, and the hostilities continued for the rest of the century. It was thus a period of tremendous political and religious upheaval—one in which regional tensions led to entanglements in much larger imperial rivalries.[10]

Within this tumultuous world, learning and knowledge had very specific definitions. For Christian scholars like Ṣagā Za'āb, formal religious education followed an extremely rigorous curriculum: after elementary instruction in languages and prayers, aspiring scholars underwent more intensive training in liturgy, scripture, music (*zéma)*, and poetry (*qené)*. The next stage was the *maṢhāfe bét* (house of the book), where students pursued advanced subjects like theology, exegesis, computus (*bāhra hasāb*), and history (*tārik)*.[11] Completing this entire educational curriculum could take decades, and though would-be scholars were often obliged to lead an itinerant life of poverty during their studies, after completing it they possessed an advanced literacy and unparalleled level of learning that distinguished them from the majority of the population. Education also uniquely prepared graduates for a range of occupations in the monastic orders, the church hierarchy, and the imperial court. Intellectuals like Ṣagā Za'āb who elected to serve the emperor as court scholars could find work as scribes, diplomats, advisors, and official historians (*Ṣahafi ta'azez*), and these bureaucratic and political posts could require further studies in specialized subjects like law and administration *(ser'āta mangest)*.

Though their careers might have varied, all these scholars preserved their knowledge in manuscripts housed in church and imperial libraries. In the sixteenth and seventeenth centuries, these parchment books and scrolls featured the scholarly language of Ge'ez, with occasional notes in Arabic and Coptic, and they often had highly eclectic contents. Some manuscripts were devoted to a single text, while others contained numerous original and copied works representing various disciplines. Still others fulfilled a documentary function, describing the intricacies of administrative protocol, land grants, or family inheritances. Generally speaking, there was little room for authorial identity in these manuscript texts: while prestige works were sometimes associated with specific individuals, as with the world history of Giyorgis Walda 'Amid (n.d.), in most cases scholars worked anonymously as compilers, copyists, or translators, only occasionally making themselves known through marginal and colophon notes. Manuscripts were rarely read outside scholarly circles, but their contents could reach a wider audience through state rituals and the religious ceremonies of the church.[12]

This intellectual world became increasingly cosmopolitan in the tumultuous sixteenth and seventeenth centuries, as is suggested by the career of one of Ṣagā Za'āb's contemporaries, the brilliant scholar 'Enbāqom (ca. 1470–ca. 1560). According to his hagiography, 'Enbāqom was born into a Jewish family in Yemen and came to highland Ethiopia in the late fifteenth century, at which time he converted to Christianity, completed a religious education, and became a monastery abbot (*'ečagé*) and advisor to the emperor Galāwdéwos (1521–1559).[13] His ecclesiastical and imperial responsibilities were many, and 'Enbāqom was also an extremely prolific scholar in dialogue with the wider world, as his writings attest. He translated several foreign texts from Arabic into Ge'ez, including a Christianized version of the life of the Buddha (*Baralām wayeswāsef*) and a canonical computational treatise by the Egyptian Copt 'Abū Šākir b. Abī l-Karam BuṢrus (n.d.),[14] and he is likely the author of a royal history that documents the arrival of the Portuguese Jesuits.[15] His linguistic abilities also tended toward the foreign and exotic: in addition to Ge'ez and Arabic, he knew Coptic, Syriac, Latin, Portuguese, and Italian—the last three acquired through his contacts with the growing number of European visitors to the Red Sea region. The implications of this remarkable career are twofold: on the one hand, 'Enbāqom's biography and writings suggest that the boundaries of Ethiopian intellectual culture became increasingly porous in the early modern period, and that at least some local scholars began to engage texts and individuals from the Mediterranean and Indian Ocean arenas. On the other hand, 'Enbāqom's career also confirms the extent to which learning, literacy, and the production of knowledge remained the domain of an intellectual elite closely linked to power. Though his personal connections, reading habits, and linguistic abilities may have crossed cultural boundaries, his allegiance to the Christian empire was clear.

Ethiopian peregrinations in the wider world fed this domestic cosmopolitanism. Local Muslims sometimes made the pilgrimage to Mecca and sought training in the Islamic sciences that was difficult to obtain in the Christian-dominated highlands, and there were permanent residences *(riwaks)* for Ethiopian Muslims in Mecca, Medina, and at Cairo's al-Azhar University.[16] The most famous of these diasporic Muslims was the Egyptian historian 'Abd al-Rahman b. Hasan al-Jabarti (1753–1825), whose major work includes a discussion of his ancestral home.[17] But Ethiopian Christians also traveled abroad throughout this period. In some cases, they left the Red Sea region to make the pilgrimage to the Holy Land, where there was a permanent diasporic presence at the monastery of Dér Seltān in Jerusalem. Some pilgrims sojourned at Coptic monasteries in Egypt, while others pressed on to Ethiopian communities in Lebanon and Cyprus.[18] It is possible that some of these Christian travelers found inspiration in the epic tales of Makédā, the Queen of Sheba, and her ancient journey from Ethiopia to Israel.[19] The example of 'ÉwosṢātéwos (ca. 1273–ca. 1352), an Ethiopian saint who visited Cairo and Jerusalem before settling among the Orthodox Christians of Armenia, could have provided a similar example.[20]

By the fifteenth and sixteenth century, developments in the wider world began to extend the range of these foreign travels. As the Ottoman conquests in the eastern Mediterranean remade the existing boundaries between Muslims, Latin Christians, and Eastern Christians, many Ethiopians fled the instability of the Holy Land and sought sanctuary in Christian Europe. Most found refuge in Italy, but others made their way further west and north.[21] By the late fifteenth century, the diasporic Ethiopian community in Rome was so numerous that Sixtus IV (1414–1484) created a permanent residence for them, the hospice that became known as Santo Stefano degli Abissini.[22] From this point onward, a steady flow of pilgrims, scholars, and ambassadors began to arrive in Rome: some came as refugees from the Ottoman lands, and others as representatives of Ethiopia's rulers, who increasingly sought to enlist Europeans in their regional conflict with the Muslims.[23]

These developments contributed to the gradual growth of an Ethiopian scholarly diaspora in early modern Europe. Some intellectuals arrived from the Eastern Mediterranean: Tomās Walda Samu'él, the monk who assisted Johannes Potken (ca. 1470–ca. 1525) in his effort to produce a Polyglot

Psalter, came to Rome from Jerusalem and lived at the Santo Stefano residence in the early 1500s. Tasfā Ṣeyon and some companions similarly arrived as refugees from the Holy Land, while the papal advisor Giovanni Battista Abissino traveled from his birthplace in Cyprus to Portugal and Spain before finally settling in Italy. Other Ethiopians came to Europe from the Red Sea region. For example, Ṣagā Za'āb's diplomatic mission took him from Ethiopia directly to Portugal, though he would have visited Rome had he not been detained in Lisbon. The scholar Gorgoryos was a different case altogether: a Catholic convert, he fled Ethiopia with the Jesuits and sought refuge first in Goa and then Rome. He briefly visited Germany while in Europe, and died while making the return voyage to Ethiopia. Though the sparseness of the documentary record makes it difficult to assess the scale of this migration, there is some evidence that a number of other Ethiopian travelers followed similar itineraries. For example, between 1519 and 1523 the Venetian Alessandro Zorzi collected a sizeable number of testimonies from Ethiopian pilgrims that describe visits to Cairo, Mount Sinai, Tunis, and Europe.[24] It would seem that the Ethiopian diaspora in Europe was small but fairly well established.

As they ventured into this wider Mediterranean world, the Ethiopians joined other cultural mediators who were then brokering new interregional flows of knowledge.[25] The best known of these are the Byzantine scholars who came to Europe following the Ottoman conquests, and who brought with them classical texts that complimented the interests of Italian humanists. Other visitors came from further to the east: Moses of Mardin (n.d.), a Syrian Orthodox scribe and Santo Stefano resident, assisted with the first printing of the Syriac New Testament and taught Syriac to several scholars in the mid-sixteenth century, and Mar Yosef (n.d.–1568), a member of the Syrian Malabar Church, obtained a number of Syriac and Persian manuscripts for the Vatican library.[26] North Africans also acted as intermediaries in this period: the best known example is the learned captive al-Hasan al-Wazzan (ca. 1494–n.d.), who lived in Rome and through his scholarship became widely known to Europeans as Leo Africanus.[27] Others include the Egyptian Copt Joseph Barbatus (ca. 1580–n.d.), who taught Hebrew and Arabic in Europe, met Thomas Erpenius (1584–1624) and Johannes Kepler (1571–1630), and worked as a manuscript collector in Constantinople, and the Moroccan merchant Ahmad b. Qasim al-Hajari (ca. 1570–n.d.), who visited scholars in Paris and Leiden.[28] Similar too were the transimperial Venetian and Ottoman dragomen who acted as cross-cultural brokers of information throughout this period.[29] All these intermediary figures helped to articulate the relationship and distinction between East and West even as older religious paradigms of accommodation and antagonism continued to shape commerce and conflict in the Mediterranean arena.[30]

The Ethiopian scholarly diaspora had much in common with these other individuals and groups. They introduced European intellectuals to new knowledge, though unlike their Muslim and Byzantine contemporaries, their scholarship did not directly engage the classical tradition that most interested European scholars. They served as ambassadors, correspondents, and translators in the growing dialogue between Europe and the wider world, much like the dragomen employed by the Ottomans and Venetians. In some respects, they resembled other African elites in Europe in that they enjoyed a freedom while abroad that was denied to most Africans in the early modern period, whether in Europe, the Americas, or the Indian Ocean arena. And like all these intermediary figures, they arrived at a moment when Europeans' understanding of the wider world was in flux.

Transmitting and Producing Ethiopian Texts

As much as these developments were the product of changing imperial dynamics, they were also tied to a host of new institutions that were then transforming Europe's intellectual landscape. State information centers like Seville's Casa de Contratación and Lisbon's Casa da Índia e Mina became hubs for the dissemination of imperial news, humanist societies like the Collège de France joined the older universities, and specialized institutions like Leiden University became centers for the study of Africa and Asia.[31] The Vatican contributed to these developments through its focus on communion with the Eastern Christians, and by the sixteenth century Rome had emerged as a major center for early orientalist scholarship. Papal emissaries to the Orthodox world acquired exotic foreign manuscripts for the Vatican library, communion-minded popes like Gregory XIII (1502–1585) surrounded themselves with specialists in Eastern Christianity and sponsored their scholarly endeavors, and new institutions like the Jesuit and Maronite Colleges, the Medici Oriental Press, and the Congregation for Propaganda Press embodied the Vatican's increasing support for the study of the wider world in the context of the Tridentine impulse toward global mission.[32]

The Santo Stefano residence was closely linked to these intellectual and institutional developments. Over the course of the sixteenth and seventeenth centuries, it became the main forum for Ethiopian contact with the wider republic of letters, and it was one of several Roman destinations for European scholars of Eastern Christianity, who were often obliged to travel given the uneven geographic distribution of knowledge about Africa and Asia.[33] The German Johannes Potken first noticed Ethiopian monks while they worshipped in the piazzas of Rome, and after this chance encounter he made inquiries that eventually led him to Santo Stefano.[34] A few decades later, Santo Stefano residents Tasfā Ṣeyon and Giovanni Battista Abissino met the Jesuits Ignatius Loyola and Francesco Alvarez (1465–1540) while they were at the Vatican.[35] Tasfā Ṣeyon also met the

French orientalist Guillaume Postel (1510–1581) while he was in Rome—they discussed the Ethiopian book of Enoch—and at some other point he was interviewed by the historian Paolo Giovio (1483–1552).[36] These Roman contacts continued in the seventeenth century, when Atanasio Kircher (1602–1680), Hiob Ludolf, and other orientalists came to Santo Stefano for extended periods of study with its resident experts.[37] Indeed, Ṣagā Za'āb's collaboration with De Góis is somewhat unique in that it is one of the few scholarly encounters that began outside the quiet halls of the Vatican residence.[38]

Personal networks and relationships were essential to these contacts between Ethiopians and Europeans, as they were more generally for scholars in this period.[39] For example, it is unlikely that Tasfā Ṣeyon could have found a printer for the *Testamentum Novum* without the assistance of his European students Pietro Paolo Gualtieri (n.d.–1572) and Mariano Vittorio (n.d.); in turn, his teachings deeply informed their own work as scholars of Eastern Christianity. But Tasfā Ṣeyon's printed work also depended on the assistance of Takla Giyorgis (n.d.), an Ethiopian scholar in Venice who secured for him a necessary Pauline epistle through his connection to the Ethiopian diasporic community in Cyprus; this item was then transported to Rome by Cardinal Alessandro Farnese (1520–1589), an influential patron of the Jesuits and scholars of Eastern Christianity.[40] And when the *Testamentum Novum* was finally completed, Farnese's grandfather Paul III (1466–1549) contributed to the project by donating money to Santo Stefano and sending copies of the work to a number of European rulers.[41] Papal patronage of this sort supported the careers of several other Ethiopian scholars: for example, Leo X (1475–1521) appointed Tomās Walda Samu'él the abbot of Santo Stefano, a position that led to his collaboration with Potken.[42] And the papal advisor Giovanni Battista Abissino received a large library for the hospice as a personal gift from Gregory XIII, another great proponent of Catholic dialogue with the wider world.[43] But papal attention did not always benefit the Ethiopians. In the early seventeenth century, Urban VIII (1568–1644) incorporated the manuscripts from the residence library into the Vatican collections, and in the mid-seventeenth century, the Santo Stefano residence was temporarily given to the Copts.[44] These examples illustrate the extent to which patrons, collaborators, and larger social networks sustained, shaped, and publicized the Ethiopians' activities.

The print medium also defined these intellectual encounters. Though the Ethiopian scholars emerged from a world of manuscript learning, they showed great interest in the possibilities of the new European technology, like several other African and Middle Eastern visitors to Europe.[45] This interest resulted in some of the earliest printed texts by non-Europeans, and certainly some of the first printed works in a Semitic language. The first printed Ge'ez work was Potken's Polyglot Psalter, produced from a Vatican manuscript through the assistance of Tomās Walda Mika'él and published by the Roman press of Marcello Silber (n.d.) in 1513, one year before the first printed Arabic work, a Vatican-sponsored translation of the Book of Hours.[46] In 1540, a Louvain press produced the Latin translation of Ṣagā Za'āb's work; it was followed in 1548 by the Roman publication of Tasfā Ṣeyon's *Testamentum Novum,* and in 1549 by a second volume containing the Pauline epistle from Cyprus and two original works by Tasfā Ṣeyon himself.[47] That same year, Tasfā Ṣeyon completed two more collaborative works describing the Ethiopian mass and baptism, and Latin translations of these were published by Antonio Blado (n.d.) in Rome.[48] After his death, Tasfā Ṣeyon's students continued to produce works that reflect his influence: Mariano Vittorio Ṣublished a Ge'ez grammar in 1552, and Pietro Paolo Gualtieri completed another Latin work on the Ethiopian liturgy.[49] By this time, the Jesuit descriptions of Northeast Africa were beginning to appear in print, and these occasionally contained Ethiopian texts in translated and/or abbreviated form. The account of Manuel de Almeida (1579–1646), for example, contained a Latin summary of *Yagalā tārik* (History of the Gallā), an ethnographic study of the Oromo by the Ethiopian scholar Bāhrey.[50] The final work with a connection to the Ethiopian diaspora was Ludolf's monumental *Historiam Aethiopicam,* published in Germany in 1681 after an extensive collaboration with Gorgoryos.[51] By the end of the seventeenth century, then, European readers had a substantial number of Ethiopian texts available to them in Ge'ez and translation, and with the exception of the Jesuit accounts, these were all linked to the early modern Ethiopian presence in Europe.

The process of producing these printed texts was not without its challenges. Obtaining the necessary Ge'ez type was both costly and difficult, just as it was for scholars and publishers interested in printing works in other exotic languages like Arabic, Syriac, Armenian, or Aramaic.[52] In his introduction to the *Testamentum Novum,* Tasfā Ṣeyon noted that it had cost fifty ounces of gold to obtain type for the more than two hundred Ge'ez characters, and that he and his colleagues had worked "day and night" for two years on the project.[53] Such a need to commission was common: Vittorio's 1552 work thanked Marcello Cervini (1501–1555), the future Marcellus II, for sponsoring the commission of Ge'ez type, and when Teseo Ambrogio produced a work of comparative linguistics in 1539, he had to obtain type for over ten languages, including Ge'ez.[54] Compounding these obstacles was the fact that individuals were occasionally possessive of their typographical investments: Potken, for example, took his Ge'ez type with him when he returned to Germany, leaving his printer with the expense and difficulty of obtaining another set if he wished to produce a second edition of the Psalter.[55] Printers of translated editions also faced this difficulty: Samuel Smith (n.d.), publisher of a 1682 English edition of Ludolf's *Historiam Aethiopicam,* chose to forego the Ge'ez text and exotic paratextual elements, instead offering readers a simple translation of the Latin text.[56] This last case illustrates how typographical issues occasionally led to major disparities in content between different editions.

Despite these challenges, the Ethiopians and their students still managed to use the print medium to introduce European readers to a new world of Christian learning. In some cases, they did this by preserving oral arguments, testimonies, and traditions in printed form. For example, Ṣagā Za'āb, who lost his manuscript collection on the voyage to Europe, nonetheless provided his readers with the first detailed discussion of Ethiopian theology in a European language. After opening his work with the traditional invocation of the trinity, he provided in-depth discussions of Christology, the importance of the Sabbath, and other more esoteric topics such as the Mariological notion of the double virgin (*dengel bakelé*).[57] Until this point, Europeans had learned of Ethiopian Christianity only through the indirect oral testimonies of other Europeans, so such detailed self-representation in text was without precedent.[58] In much the same way, Ludolf's work documented the oral testimony of his collaborator Gorgoryos, who discussed such diverse topics as Ethiopian astronomy, the gradual spread of the Amharic language, and Christian perceptions of the Muslims in the sixteenth century.[59] Other printed works preserved a wealth of Ethiopian oral knowledge about the ethnic groups, languages, geography, political institutions, and legal systems of the region: Potken, Tomās Walda Samu'él, and Vittorio all provided detailed outlines of Ge'ez grammar; Ṣagā Za'ab described royal succession and the relationship between the Ethiopian and Egyptian church hierarchies; and Ludolf and Gorgoryos documented the place names of the region in extensive detail.

In other cases, Ge'ez printed works went beyond oral knowledge to introduce readers to actual texts from Ethiopia's scholarly tradition. Tasfā Ṣeyon's publications reproduced a number of religious works that were previously available only in manuscript form: most notably, his *Testamentum Novum* included an Ethiopian recension of the Gospels, epistles previously unknown to Latin Christians, and examples of Ethiopian Apostolic and Marian anaphora; he also published the Ethiopian Missal. Other printed editions transmitted somewhat less sacred texts to European readers. Ludolf's work includes examples of several genres of traditional Ethiopian historical writing: a king list (*tārik negest*), two prosopographical lists of the Alexandrian and Ethiopian patriarchs, and a hagiography (*gadl*) of the Ethiopian saint Takla Hāymānot (n.d.–ca. 1313).[60] Equally significant was Ṣagā Za'āb's summary of the *Kebre negest* epic, perhaps the most prestigious text in the Ethiopian canon. All these works were essentially unknown in Europe at that time, so for European readers who had not been to Ethiopia, these would have been the first glimpses of the region's remarkable literary tradition.

In a few cases, the printed works also introduced readers to features of Ethiopian intellectual culture. With the assistance of Gorgoryos, Ludolf sketched the canon of Ethiopian scholars and noted the names of some preeminent figures whose works he had yet to consult.[61] He also included several discussions of literacy patterns and educational institutions, and observed that in Ethiopia scribes could be paid to produce texts.[62] Similarly, Ṣagā Za'āb provided a description of major works and figures in the canon, and he explicitly noted how these informed his oral knowledge:[63] for example, he noted that his testimony on "the discipline, doctrine, and law" was based on what "the Apostles in their holy books of councels and canons . . . have taught us," and on those "books of the ordinance of the Church [of which] there be eight, all which were complied by the Apostles when they were assembled together at Jerusalem."[64] He was thus representing an authoritative intellectual tradition, not just his own musings.

Taken together, these diverse works offered European readers a sophisticated introduction to Ethiopia's oral knowledge, scholarly tradition, and intellectual culture. In the first decades of the sixteenth century, Potken had been struck by the great difficulty of communicating with the exotic Ethiopian pilgrims of Rome, and deeply frustrated by the lack of information about them and the great challenge of obtaining it.[65] Less than one hundred years later, Ethiopian scholars and their European colleagues had begun to make the languages, texts, and ideas of the Red Sea region widely available to European readers through the new medium of print.

Lost in Translation?

But what were the intellectual consequences of this interregional flow of knowledge? We can answer this question by considering some of the currents that framed European readings of Ethiopian works. By the sixteenth century, some European scholars had begun to focus their attention on the exotic people, cultures, and places of Africa and Asia in new ways. Early Renaissance trends disparaging the study of nonclassical subjects like the Muslim world had yielded to new interests in Kabbalah and Hebraism, the exotic writings of Eastern Christianity, and the production of comparative Polyglot Bibles that juxtaposed scripture in different languages on a single page.[66] Among Catholics, these outward-looking trends were strengthened by Tridentine emphases upon the harmonization of faith, communion with Eastern Christians, and renewal through global mission, developments that were embodied in Rome's new institutional landscape. And across the confessional divide, a few Protestant scholars began to see new value to the study of Eastern Christianity. For example, the English scholars William Bedwell and Michael Geddes each believed that the history of Orthodox Christianity could guide their own attempts to create independent churches in confrontation with Catholicism—after all, as Geddes noted, this was a tradition "that was never at any time under the Papal yoke"—and for this reason they turned to the study of the Arab world and Ethiopia, respectively.[67] Historian Robert Irwin argues that all these intellectual currents contributed to the emergence of early orientalism.[68]

Equally significant was the growing preoccupation with geographic and cultural discovery. Thinking broadly about the early modern world, Sanjay Subrahmanyam has observed that this was a period in which "cultures of travel" proliferated on a global scale, as people in various regions recognized the limitations upon their respective known worlds and actively sought to expand and redefine these.[69] The most obvious example of this trend is the European maritime exploration of the sixteenth and seventeenth century, which led to the post-Columbian interest in the world across the Atlantic, and following the dramatic crossing of the Pacific Ocean, the intensification of contacts between Europe and East Asia. However, as Subrahmanyam points out, the early modern preoccupation with discovery also manifested itself in the flourishing of travel writing and protoethnography. Though exponents of these genres can be found in many world regions, in the European context this was the era of figures like Giovanni Battista Ramusio (1485–1557), Richard Hakluyt (1552–1616), and Samuel Purchas (1577–1626), who compiled and published massive volumes of travelers' descriptions of all that was novel and exotic, and of scholars like Bartolomé de Las Casas (1484–1566) and José de Acosta (1539–1600), who attempted to interpret and explain these differences for readers.[70] Historians have differed in their assessments the intellectual consequences of these developments.[71]

Given this sixteenth-century convergence of orientalism and cultures of travel, we can say that the Ethiopians scholars arrived at a moment of intense European preoccupation with the wider world and its implications. It is thus unsurprising that they became authoritative sources for the growing number of readers interested in Africa, Asia, and Eastern Christianity, and that their works were regularly cited and reprinted throughout sixteenth and seventeenth centuries. Ṣagā Za'āb was particularly influential, and his work remained a canonical text for nearly two centuries despite its censorship by the Catholic Church. Erasmus learned of him through his student De Góis, and he indirectly referred to him on at least one occasion;[72] an appropriately retitled version of his work (*The Errors of the Ethiopians*) circulated among Jesuits throughout the sixteenth century;[73] and in 1611, an English edition of it appeared.[74] By then, scholars were routinely citing Ṣagā Za'ab in a variety of contexts: Morton Eudes (n.d.) and Edward Brerewood (1565–1613) mentioned him by name in their respective theological and historical studies, and Ludolf openly used Ṣagā Za'ab's work to corroborate the oral testimony of Gorgoryos.[75] Though this was the best known text by an Ethiopian author, other works also enjoyed reprinting and translation. Tasfā Ṣeyon's *Testamentum Novum* was likely the source of the Ge'ez text that Giovanni Battista Raimondi (ca. 1584–1614) hoped to include in his Polyglot Bible in the late sixteenth century, and the Jesuit Nicolao Godinho (1561–1616) mentioned him (and Ṣagā Za'ab) by name in an eclectic 1615 compilation.[76] In 1657, Brian Walton (1600–1661) excerpted the *Testamentum Novum* in his Polyglot Bible, adding to it Potken's work as well.[77] The work of Tasfā Ṣeyon's student Vittorio also went through multiple editions: the Congregation of Propaganda Press produced a version of it in 1630.[78] Ludolf's work was similarly reprinted and translated throughout the seventeenth century, and it became a definitive reference on Northeast Africa from the eighteenth century onward.[79]

This dissemination through print led some diasporic Ethiopian scholars to an authorial identity and wide readership that would have been impossible in Christian Ethiopia's elite world of learning and scholarship. Ṣagā Za'āb and Gorgoryos became closely associated with the texts of De Góis and Ludolf, and their motives, honesty, and intelligence even became subjects of academic controversy among European specialists.[80] However, this kind of fame was not guaranteed: Tasfā Ṣeyon, easily the most prolific and consulted of the Ethiopian scholars in early modern Europe, remained relatively obscure outside the community at Santo Stefano, and his Latin works on the Ethiopian liturgy do not appear to have been reprinted, despite their potential value for Catholic readers interested in global mission or communion with Eastern Christians.

However, while the Ethiopians' works were widely read and cited, they were only slowly and unevenly reconciled with the existing canon of knowledge, much like the new information then arriving from Asia and the Americas. In some cases, the new printed works explicitly noted problems with older references: for example, Ṣagā Za'āb critiqued an earlier description of Ethiopia by an Armenian named Matéwos (n.d.), and Ludolf explained where information provided by Gorgoryos challenged the testimony of other writers.[81] But even with these explicit interventions, many European readers continued to be influenced by classical and medieval ideas about Africa, despite the availability of new and seemingly more accurate sources of information. For example, a number of scholars continued to erroneously refer to Ge'ez as Chaldaean: in 1513, Potken himself made the mistake with his Psalter, and Teseo Ambrogio employed it in his 1539 work.[82] One year later, De Góis used the misleading term in his collection, and the 1552 and 1630 editions of Vittorio's grammar repeated the mistake.[83] By the seventeenth century the continued use of the archaic term had become so anachronistic that Ludolf remarked on it in his work, adding that "neither had Gregory [Gorgoryos] ever heard it in his own country."[84] Several other outmoded medieval notions continued to remain influential. The idea of Prester John, for example, persisted for centuries: though Ṣagā Za'āb and Ludolf both explained the confusion surrounding the term, several European scholars continued to use it in reference to the Ethiopian emperor.[85] And as late as 1693, one scholar even continued to employ the classical meaning of the term Ethiopia, as a general reference to Africa.[86] These various semantic confusions highlight the extent to which classical and medieval frameworks continued to inform the early modern European imagination of the wider world, even when new works explicitly refuted earlier claims. In this respect, European conceptions of Africa and Asia were similar to their understanding of the Americas: recent discoveries did not immediately eradicate old ideas, but instead accumulated alongside them.[87]

In addition to the established canon, translation also mediated Europeans' reception of the Ethiopian scholarship. Language was one dimension of this process: in general, it seems that translated texts (Ṣagā Za'āb) became widely known through multiple editions, while untranslated ones (Tasfā Ṣeyon) did not. This likely reflects the fact that very few Europeans could read Ge'ez in the early modern period, and those who could were most probably abroad.[88] But other acts of translation also informed how Europeans understood the Ethiopians' texts. Ludolf obviously mediated the oral testimony of Gorgoryos, and De Góis did the same as he translated and edited Ṣagā Za'āb's work for publication. However, there is also some evidence that Ethiopians' foreknowledge of European readers periodically conditioned what they chose to say. For example, when Ludolf asked Gorgoryos about the controversial subjects of female circumcision and magical scrolls, his informant declined to comment.[89] Printers also performed acts of mediation and translation. For example, the anonymous printer of the *Testamentum Novum* shaped readers' understanding through the juxtaposition of familiar images and unfamiliar words: he included European-style woodcut illustrations alongside the Ethiopian text, and he employed paratextual markers from the European manuscript tradition, though these have a passing resemblance to their Ethiopian counterparts.[90] And as we have seen, the process of reprinting also shaped textual meaning: for example, the comparative prerogatives of Polyglot Bible scholarship meant that later versions of *Novum Testamentum* stripped the text of the Anaphora that Tasfā Ṣeyon had deliberately included in his original edition, thereby repackaging his work in a less exotic but more useful form.

Thus a complex matrix of factors shaped Europeans' understanding of these works. But it is also likely that readers were largely unaware of the intended purpose and rhetorical context of some of the Ethiopians' texts. Reflecting on the tumult of the sixteenth-century Northeast Africa, the royal historian of the emperor Galāwdéwos noted the two "great worries" that defined his lifetime: on the one hand, the simmering regional conflict between Muslims and Christians; and on the other, the many "controversies with the learned Franks *('aferenṢ)* pertaining to their fear of faith."[91] The second point is key: while the anti-Christian jihad of the Adal Sultanate was a cause for alarm among Ethiopian Christians, the growing tensions with their European coreligionists were also a significant and pressing concern. After the arrival of the Portuguese Jesuits in 1520, the Ethiopian church faced a spiritual challenge from foreigners who "critique[d] the true faith that had been sent from Alexandria" and "loudly proclaim[ed] the distorted belief issued in Rome," as our royal historian put it.[92] He continued by noting that in response to European missionary activities in the region, Ethiopian scholars began to assemble "a great number of treatises" that critiqued Catholic doctrine and defended Orthodoxy.[93] Some of these works are known to us today: they include original texts like *Sawen nafes* (Refuge of the Soul) and *Mazgab hāymānot* (Treasury of Faith), and also translated Arabic texts from the Orthodox traditions of the Middle East, such as the patristic work *Hāymānot 'abaw* (Faith of the Fathers).[94] This period also saw the appearance of the Ge'ez epic of Saint Tertag, a seemingly allegorical tale of an Armenian ruler who severed his religious ties with Rome.[95] Even the emperor Galāwdéwos himself wrote a learned defense of Orthodoxy and gave it to the Portuguese missionaries, who returned it to Europe where it was translated and published by Ludolf in 1661.[96] The Ethiopian scholarly diaspora thus emerged from a world filled with religious conflicts on several fronts, conflicts both of arms and polemic—the first between the Muslims and Christians of Northeast Africa, and the second between European and Ethiopian Christians.

There evidence that the "great worries" of our royal historian informed at least two of the Ethiopians' scholarly activities in Europe. Ṣagā Za'āb clearly wrote within the rhetorical context of the larger inter-Christian debate between Ethiopians and Europeans; indeed, though he was impelled to write by his hostile Portuguese interlocutors, instructive scholarship, he claimed, was a central purpose of his visit. He explained in the introduction to his work that his patron Lebna Dengel had commanded him to write so that "the Godly Christians of Europe may understand our customs and the integrity of our manners."[97] If we bear in mind the inter-Christian rivalries then at work in Northeast Africa, it would seem that Ṣagā Za'āb shared the same intellectual urge as the anti-Catholic translators and polemicists in Ethiopia, despite the fact that their works were dispersed across tremendous geographic distance and took different textual forms. Like his counterparts at home, he intended to defend his faith through rigorous exposition and the documentation of European error. It was simply the approaches that differed: Ṣagā Za'āb defended Orthodox Christianity to European readers, in translation and through the medium of print.[98]

Tasfā Ṣeyon's work appears to reflect the other "great worry" of his day, the Muslim-Christian conflicts then engulfing Northeast Africa. In his study of Tasfā Ṣeyon's correspondence, Ignazio Guidi suggests that since many of Ethiopia's church manuscripts had been destroyed in the era's violent conflicts, Tasfā Ṣeyon may have seen the production of printed texts as a remedy to this catastrophe.[99] Though Guidi does not offer evidence for this pastoral purpose, some features of the *Testamentum Novum* do support this interpretation. First, Tasfā Ṣeyon's presentation of the Gospels curiously features a European astronomical table labeled in Ge'ez.[100] This addition suggests an imagined Ethiopian reader, for whom this presentation would be an introduction to European ideas about the nature of the heavens—a European reader would have likely found such content out of place in an exotic Eastern Christian text. Second, Tasfā Ṣeyon's print edition followed the common Ethiopian practice of introducing the Gospels with the Eusebian canon tables, again suggesting that he had a specifically Ethiopian mode of reading in mind as he prepared his work.[101] His text, these tables suggest, was intended for the same use as any other Ethiopian Gospel—it was to be read as scripture, not as an academic or linguistic resource.[102] Yet later reprints of the *Testamentum Novum* removed precisely these elements—the instructive illustrations and the useful reading guides—in order to include it in Polyglot Bibles, effectively repackaging the Ethiopian scriptures alongside their Arabic, Syriac, Latin, and Hebrew counterparts. Thus the interests of European scholarly readers increasingly shaped subsequent editions of Tasfā Ṣeyon's work, and their goal of accessing greater scriptural truth through comparative analysis gradually effaced the Ethiopian's original pastoral purpose. His intentions, it seems, were lost in translation.

Conclusion: Globalization, Fantasy, and Philology

Our story of travel, scholarship, and transmission through print has some larger implications for our understanding of cosmogony in the early modern world, but to grasp these we need to think broadly about cross-cultural intellectual exchange. A number of historians have recently begun to analyze interregional links in this period through the framework of archaic globalization.[103] In their view, the ties of the early modern world were more multifaceted than is sometimes supposed, for they emerged from common cultural patterns as well as long distance trade. Universal religion, cosmic kingship, and tastes for exotic consumption forged ongoing interregional relationships throughout Afro-Eurasia, and though these were not the proper global ties produced by the maritime exploration of the sixteenth century, these older relationships were just as multicentered and dynamic as their successors. Crucially, Anthony Hopkins and C.A. Bayly also suggest that while later forms of globalization relied upon new technologies, information networks, and imperial systems, they also depended upon the subordination and "cannibalization" of these archaic ties of culture and trade.

The story of the early modern Ethiopian scholarly diaspora in Europe neatly illustrates this argument about the links between older and newer forms of globalization, and more to the point, it shows how the interface between these informed world-building projects in this period. In the sixteenth and seventeenth centuries, European constructions of Africa and Asia clearly depended upon a diverse set of agents, institutional networks, and interregional ties, and the origins of these ranged widely. On the one hand, contacts between Europeans and Ethiopians in this period emerged from a number of interrelated but still ultimately regional—and hence archaic—processes: the instability that followed Ottoman imperial expansion in the Eastern Mediterranean, the Muslim-Christian conflicts in Northeast Africa that led the divinely sanctioned rulers of Ethiopia to seek diplomatic ties with their European coreligionists, and the ultimately ecumenical sources of European scholarly interest in Eastern Christianity. Yet on the other hand, the transmission of Ethiopian knowledge to European readers also relied on some decidedly new developments: the print technology and institutional patronage that sponsored the flow of oral and written knowledge from Ethiopia, and the intellectual and institutional framework of global salvific Catholicism that sustained a new interest in distant people and places.[104] No less important were the reverberations of the new imperial dynamics at play: just as the Ottoman conquest of Jerusalem inspired some Europeans to pursue divinely ordained voyages of maritime exploration, so too did it push pious Ethiopian Christians to seek refuge across the Mediterranean.[105] Christian universalism bound these new and old components together, creating a conjuncture for ecumenical intellectual exchange even as tensions between Catholics, Protestants, and Orthodox mounted. Globalization, and in particular the overlap between its archaic and protomodern forms, was at the heart of this intellectual exchange between the Red Sea and the Mediterranean.

An entirely new conceptualization of the wider world was the result, for the products of this early modern exchange directly contributed to the emergence of the orientalism—an intellectual revolution with far-reaching results. Because of the publications of the sixteenth century, European scholars of the seventeenth century began to understand Africa and Asia through a new and highly specialized philological lens, and through it they reconceptualized the diverse societies of these regions. Ethiopian scholars contributed to this process by unintentionally providing a canonical set of texts that became standard references in the coming centuries. Their expertise was cited and their work read by illustrious scholars like Erasmus, Postel, Loyola, and Kircher, and their intellectual influence continued long after their death through the extensive reprinting of their writings. As a result, the European understanding of Africa was for the first time directly informed by African forms of knowledge. Seen in this light, these Ethiopian texts and the commentary they produced represent a turning point in the European study of Africa, the Middle East, and Eastern Christianity: they exemplify the shift from fantasy to philology that underpinned early orientalism, and they underpinned the emergence of semitic studies and its Northeast African branch.[106] At the same time, these works—together with the writings of al-Hasan al-Wazzan—represent the formal introduction of African studies in Europe, though this field would not again find an institutional basis until the nineteenth century.

Yet this was an ultimately unequal exchange. While Europeans acquired an increasing amount of knowledge about the Red Sea region, there was no reciprocal flow of information. Instead, contact with Europeans led many Ethiopian Christian scholars to focus ever more intently on the world of their Orthodox coreligionists in the Middle East. As they broadened their canon and intellectual horizons in an attempt to make sense of their changing world, they turned increasingly to the familiar, for the most part refusing the European exports of the Jesuits. Indeed, one of the few early modern Ge'ez translations of a European text, *Zéna ʿāyehud* (History of the Jews), was in fact made from an Arabic version likely obtained through the Ethiopian diaspora in the Eastern Mediterranean.[107] In the same way, artistic and architectural borrowing in the Red Sea region remained rooted in the archaic networks and trade diasporas of the Indian Ocean arena, though an interest in European iconography remained after the departure of the Portuguese Jesuits. In these respects, early modern Ethiopia had much in common with East Asia in the same period: as Geoffrey Gunn suggests, the latter also participated in an unequal exchange of knowledge with Europe.[108] It is telling that in both regions, cultural creolization did not occur until the nineteenth century, despite these much earlier precedents of intellectual contacts.

But is such intellectual and cultural refusal an act of world building? We can conclude by returning to the work of Ṣagā Za'āb, which offers one possible answer to this question. Facing his hostile Catholic interlocutors, and writing far from his home and colleagues, he proposed to his readers a new conception of inter-Christian dialogue, a plea for true ecumenical fraternity that we can read as a commentary from the older world of the universal upon the emerging hierarchical mindset of the new:

> It is very unworthily done to reprehend strangers that bee Christians so sharply and bitterly... but it shal be better and more standing with wisdome, to sustaine such Christians whether they bee Greeks, Americans, or Aethiopians... for we bee al the sons of baptisme, and joyne together in opinion concerning the true faith... there is no cause why wee should contend so bitterly touching [with regard to] ceremonies, but that each one should observe his owne ceremonies, without the hatred, rayling, or inveighing of other[s].[109]

His plea was ignored.

TESTAMENTVM
NOVVM CVM EPISTOLA PAVLI AD
Hebreos tantum, cum concordantijs Euangelistarum Eusebij & numeratione omnium verborum eorundem.
Missale cum benedictione incensi ceræ et c. Alpha-
betum in lingua ግእዝ:gheez, idest libera quia a
nulla alia originem duxit, & vulgo dicitur
Chaldea, Quæ omnia Fr Petrus Ethyops
auxilio piorum sedente Paulo.III.Pont.
Max. & Claudio illius regni Im-
peratore imprimi curauit.

ANNO SALVTIS M. D. XLVIII.

Figure 8.1 Excerpt from *Testamentum Novum*.

EVANGELIVM. MATHEI.

Figure 8.2 Excerpt from *Testamentum Novum*.

Notes

I would like to thank Allison Kavey, Heather Sharkey, Matteo Salvadore, and the anonymous reviewers at Palgrave Macmillan for their comments on earlier versions of the chapter. I also thank the staff of the New York Public Library Special Collections and the Robert Hess Special Collection at the Brooklyn College Library.

1. Damião de Góis, ed. *Fides religio moresque Aethiopum sub imperio Preciosi Ioannis dentium* (Louvain: Rescius, 1540). Ṣagā Za'āb's original manuscript has been lost: his only other known work is presented in René Basset, "Deux lettres éthiopiennes du xvi siecle," *Giornale della Societa asiatica italiana* 3 (1889): 58–79.
2. Tasfā Ṣeyon, ed. *Testamentum Novum cum epistola Pauli ad Hebreos tantum, cum concordantijs evangelistarum Eusebij & numeratione omnium verborum eorundem* (Rome: 1548).
3. Renato Lefevre, *L'Etiopia nella stampa del primo cinquecento* (Como: Pietro Cairoli, 1966), 23; Hiob Ludolf, *Historia Aethiopica* (Frankfurt am Main: Martini Jacqueti, 1681) and *Sciagraphia historiae Aethiopicae* (Jena: 1676).
4. Renato Lefevre, "Documenti e notizie su Tasfā Ṣeyon e la sua attività romana nel sec. XVI," *Rassegna di studi etiopici* 24 (1969–1970): 75, 81.
5. See, for example, Stuart Schwartz, ed. *Implicit Understandings: Observing, Reporting, and Reflecting on the Encounters between Europeans and Other Peoples in the Early Modern Era* (New York: Cambridge University Press, 1994) and Anthony Pagden, ed. *Facing Each Other: The World's Perception of Europe and Europe's Perception of the World,* 2 vols. (Ashgate: Aldershot, 2000).
6. Mauro Da Leonessa, *S. Stefano dei Maggiore o degli Abissini* (Rome: Vatican, 1928) and Renato Lefevre, "Riflessi etiopici nell cultura europea del medioevo e del rinascimento," *Annali Lateranensi* 11 (1947): 255–342.
7. Cf. Kate Lowe and T.F. Earle, eds. *Black Africans in Renaissance Europe* (Cambridge: Cambridge University, 2005), Gerald Maclean, *Re-Orienting the Renaissance: Cultural Exchanges with the East* (Houndsmill: Palgrave Macmillan, 2005), and David Northrup, *Africa's Discovery of Europe, 1450–1850* (New York: Oxford University Press, 2002), which includes a discussion of Tasfā Ṣeyon. For a short but stimulating examination of the Ethiopian diaspora from the perspective of Ethiopian intellectual history, see Aleme Eshete, "Baqademo zamanāt ka1889 'āmet meheret bafit weç 'agar yatamāri 'ityoṢyānoč tārik," *Ethiopian Journal of Education* 6, no. 1 (1973): 115–148. Ethiopian-European contacts are also briefly discussed in Alastair Hamilton, "Eastern Churches and Western Scholarship," in Anthony Grafton, ed. *Rome Reborn: The Vatican Library and Renaissance Culture* (New Haven: Yale University, 1993), 225–249.
8. Though briefly noted in Lucien Febvre and Henri-Jean Martin, *The Coming of the Book: The Impact of Printing 1450–1800,* trans. David Gerard (London: Verso, 1997), 213, the only study is Hendrik Wijnman, "An Outline of the Development of Ethiopian Typography in Europe," in Hendrik Wijnman, ed. *Books on the Orient* (Leiden: Brill, 1960), 9–38. Ethiopian specialists have focused on the history of print in later periods: Stephen Wright, *Ethiopian Incunabula* (Addis Ababa: 1967), Bahru Zewde, *Pioneers of Change in Ethiopia: The Reformist Intellectuals of the Early Twentieth Century* (Oxford: James Currey Press, 2002), and Richard Pankhurst, "The Foundations of Education, Printing, Newspapers, Book Production, Libraries, and Literacy in Ethiopia," *Ethiopia Observer* 6, no. 3 (1962): 241–292.

9. Fernand Braudel, *The Mediterranean and the Mediterranean World in the Age of Philip II*, trans. Siân Reynolds (New York: Harper Colophon, 1973), 2:767–768.
10. Overviews of these developments include Taddesse Tamrat, *Church and State in Ethiopia, 1270–1527* (Oxford: Oxford University, 1972), Mordechai Abir, *Ethiopia and the Red Sea: The Rise and Decline of the Solomonic Dynasty and Muslim-European Rivalry in the Region* (London: Frank Cass, 1980), and Haggai Erlich, *Ethiopia and the Middle East* (Boulder: Lynne Rienner, 1994). On patterns of Muslim-Christian relations in the region, see Haggai Erlich, *Islam and Christianity in the Horn of Africa: Somalia, Ethiopia, and Sudan* (Boulder: Lynne Rienner, 2010).
11. The literature on Ethiopian Christian education and scholarship is immense: *Encyclopaedia Aethiopica,* ed. Siegbert Uhlig (Wien: Harrassowitz Verlag, 2003–2008), hereafter *EA,* introduces all the institutions and disciplines. See also Imbakom Kalewold, *Traditional Ethiopian Church Education,* trans. Menghestu Lemma (New York: Teacher's College Press, 1970); Paulos Milkias, "Traditional Institutions and Traditional Elites: The Role of Education in the Ethiopian Body-Politic," *African Studies Review* 19, no. 3 (1976): 79–93; and more generally, Donald Crummey, *Land and Society in the Christian Kingdom of Ethiopia* (Urbana-Champagne: University of Illinois, 2000).
12. An overview of Ethiopian manuscript culture is provided by Sergew Hable Selassie, *Bookmaking in Ethiopia* (Leiden: Author, 1981). See also the manuscripts and archives forum in *Northeast African Studies* 11, no. 1 (2010), special issue on "Genealogies of Knowledge."
13. Biographical details and a complete list of his translations and original works can be found in *EA,* 1:280–282.
14. Otto Neugebauer, *Abu Shaker's "Chronography": A Treatise of the 13th Century on Chronological, Calendrical, and Astronomical Materials, written by a Christian Arab, Preserved in Ethiopic* (Vienna: Osterreichischen Akademie der Wissenschaften, 1988).
15. 'Enbāqom cannot conclusively be identified as the author of the royal history of Galawdéwos. However, his association with the emperor and the Arabic loanwords and meticulous calendrical notation displayed in the text suggest him as the probable author: Anonymous, *Chronique de Galâwdêwos (Claudius), Roi d'Éthiopie,* ed. William Conzelman (Paris: Librairie Émile Bouillon, 1895), viii.
16. Edward Ullendorf, "Djabart," in P. Bearman et al. *Encyclopaedia of Islam* (Leiden: Brill, 2008), 2:354; David Ayalon, "al-D̲jabartī, 'Abd al-RaṢmān b. Ṣasan," *Encyclopaedia of Islam,* 2:354; and Abir, 11.
17. On al-Jabarti's ancestry, see David Ayalon, "The Historian al-Jabarti and His Background," *Bulletin of the School of Oriental and African Studies* 23, no. 2 (1960): 217–249.
18. Several such journeys were described by the Venetian scholar Alessandro Zorzi (ca. 1470–n.d.), who learned of them from interviews with Ethiopian informants. See O.G.S. Crawford, ed. *Ethiopian Itineraries circa 1400–1524*

(Cambridge: Hakluyt Society, 1958). On the diasporic communities of Lebanon and Cyprus, see Enrico Cerulli, *Etiopi in Palestina* (Rome: Liveria del Stato, 1943), 1:325–334 and 2:1–11; and Lefevre, "Riflessi etiopici," 277–302.

19. On the role of the *Kebra nagast* in later Ethiopian travel narratives, see James De Lorenzi, "Printed Words, Imperial Journeys, and Global Scholars: Historiography and Cosmopolitanism in the Red Sea World, 1800–1935," Ph.D. dissertation, University of Pennsylvania, 2008, 123–164.
20. *AE*, 2:469–472.
21. For a discussion, see Cerulli, *Etiopi in Palestina*, 1:397. An argument for the changing religious dynamic can be found in Molly Greene, *A Shared World: Christians and Muslims in the Early Modern Mediterranean* (Princeton: Princeton University, 2000).
22. Da Leonessa, *S. Stefano*, 171–177; Lefevre, "Riflessi," 260–263.
23. For an overview of these contacts, see Matteo Salvadore, "The Ethiopian Age of Exploration, 1306–1458," *Journal of World History* (forthcoming). African diplomatic missions are surveyed in Kate Lowe, "Representing Africa: Ambassadors and Princes from Christian Africa to Renaissance Italy and Portugal, 1402–1608," *Transactions of the Royal Historical Society* 17 (2007): 101–128.
24. These are surveyed in Crawford, *Ethiopian Itineraries*, ix–xiii.
25. I borrow the term "cultural mediator" in this context from E. Natalie Rothman, "Interpreting Dragomans: Boundaries and Crossings in the Early Modern Mediterranean," *Comparative Studies in Society and History* 51, no. 4 (2009): 771–800.
26. Robert J. Wilkinson, *Orientalism, Aramaic, and Kabbalah in the Catholic Reformation: The First Printing of the Syriac New Testament* (Leiden: Brill, 2007), esp. 63–94; Hamilton, "Eastern Christians," 239, 243.
27. Natalie Zemon Davis, *Trickster Travels: A Sixteenth Century Muslim between Worlds* (New York: Hill and Wang, 2006).
28. Alastair Hamilton, "An Egyptian Traveler in the Republic of Letters: Joseph Barbatus or Abudacnus the Copt," *Journal of the Warburg and Courtauld Institutes* 57 (1994): 123–150; and Gerar Wiegers, "A Life between Europe and the Maghrib: The Writings and Travels of Ahmad b. Qâsim al-Hajarî al-Andalusî," in Geert Jan van Gelder and Ed de Moor, eds. *The Middle East and Europe: Encounters and Exchanges* (Amsterdam: Rodopi, 1992), 87–115. See also the related example of Juan Latino of Granada (n.d.–1590), less a knowledge broker than an assimilated African scholar: Baltasar Fra-Molinero, "Juan Latino and his Racial Difference," in *Black Africans*, 326–344.
29. Suraiya Faroqhi, *The Ottoman Empire and the World Around It* (London: I.B. Tauris, 2007), Daniel Goffman, *The Ottoman Empire and Early Modern Europe* (Cambridge: Cambridge University, 2002), and Natalie Rothman, "Between Venice and Istanbul: Trans-Imperial Subjects and Cultural Mediation in the Early Modern Mediterranean," Ph.D. dissertation, University of Michigan, 2006.

30. On the persistence of older paradigms, see Molly Greene, "Beyond the Northern Invasion: The Mediterranean in the Seventeenth Century," *Past and Present* 174 (2002): 43–70.
31. Peter Burke, *A Social History of Knowledge from Gutenberg to Diderot* (Cambridge: Blackwell, 2000), 32–52.
32. Robert Irwin, *Dangerous Knowledge: Orientalism and Its Discontents* (New York: Overlook, 2006), 50–81; Hamilton, "Eastern Christians," 225–240.
33. Describing early orientalism, Robert Wilkinson notes "a native teacher was a rare prize, a manuscript a treasure." Wilkinson, *Orientalism,* 190, and more generally, Burke, *Social History of Knowledge,* 53.
34. Lefevre, *L'Etiopia,* 15–23.
35. Lefevre, "Riflessi," 270 and "Documenti," 81.
36. Lefevre, "Documenti," 87, 105; F. Gallina, "Iscrizioni etiopiche ed arabe di S. Stefano dei Mori," *Archivio della società romana di storia patria* 11 (1888): 287–288; Ignazio Guidi, "La prima stampa del Nuovo Testamento in etiopico, fatta in Roma nel 1548–1549," *Archivio della reale società romana di storia patria* 9 (1886): 273–278.
37. Da Leonessa, *S. Stefano,* 262.
38. For De Góis's own account of meeting Ṣagā Za'āb, see Johannes Boemys, ed. *Manners, Lawes, and Customes of All Nations, Collected out of the best Writers by Ioannes Boemys Aubanus, a Dutchman, with many other things of same Argument* (London: Francis Burton, 1611), 544–545. All references to and quotations from Ṣagā Za'āb are taken from this English translation. Discussions of the de Góis-Ṣagā Za'āb connection include Elizabeth Feist Hirsch, *Damião de Gois: The Life and Thought of a Portuguese Humanist, 1502–1574* (Antwerp: Martinus Nijhoff, 1967); Jeremy Lawrance, "The Middle Indies: Damião de Góis on Prester John and the Ethiopians," *Renaissance Studies* 6, nos. 3–4 (1992): 316–324; and Andreu Martínez, "Paul and the Other: The Portuguese Debate on the Circumcision of the Ethiopians," in Verena Boll, Steven Kaplan, Andreu Martínez, and Evgenia Sokolinskaia, eds. *Ethiopia and the Missions: Historical and Anthropological Insights* (Muenster: Lit Verlag, 2005), 53–62.
39. Irwin, *Dangerous Knowledge,* 85.
40. Their correspondence is published in Guidi, "La prima stampa del Nuovo Testamento in etiopico," 276–277.
41. Lefevre, "Riflessi," 266.
42. Ibid., 264.
43. Ibid., 268.
44. Da Leonessa, *S. Stefano,* 269–300.
45. For an account of a similar interest among Syrian Orthodox Christians, see Wilkinson, *Orientalism,* 63–64. In contrast, Natalie Zemon Davis argues that al-Hasan al-Wazzan had little interest in printing: Davis, *Trickster Travels,* 122–124.
46. The work is Johannes Potken, ed. *Psalterium in quatuor linguis hebraea graeca chaldaea latina* (Rome: Coloniae, 1513); discussions include Lefevre,

"Notizie," 98 and Wijnman, *Books on the Orient,* xv. On the Silber press, see Alberto Tinto, *Gli annali tipografici di Eucario e Marcello Silber (1501–1527)* (Florence: Olschki Editore, 1983); on the Arabic Book of Hours, see Irwin, *Dangerous Knowledge,* 75.

47. De Góis, ed. *Fides* and Tasfā Ṣeyon, ed. *Testamentum Novum.* Details on the obscure 1549 publication can be found in Giuseppe Fumagalli, *Bibliografia etiopica* (Milan: Ulrico Hoepli, 1893), 131; see also Lefevre, "Notizie," 88.
48. Wijnman, *Books on the Orient,* xix.
49. Mariano Vittorio, *Chaldeae seu Aethiopicae linguae institutions* (Rome: 1552); Gualtieri's work is described in Da Leonessa, *S. Stefano,* 199.
50. Discussions of this work include Sevir Chernetsov, "The 'History of the Gallas' and Death of Za-Dengel, King of Ethiopia (1603–1604)," in *IV Congresso Internazionale di Studi Etiopici* (Rome: Accademia Nazionale dei Lincei, 1974), 803–808; and Getatchew Haile, *Ya'abbā Bāhrey dersetoč oromočen kamimalekatu léloč sendoč gārā* (Collegeville: Author, 2002). See also Hervé Pennec, *Des jésuites au royaume du Prêtre Jean (Éthiopie)* (Paris: Centre Culturel Calouste Gulbenkian, 2003), 241–307; and Leonardo Cohen, "The Jesuit Missionary as Translator," in *Ethiopia and the Missions,* 7–30.
51. Ludolf, *Historiam Aethiopicam.*
52. On the difficulties faced by European printers of Arabic and Syriac texts, see Hamilton, *William Bedwell,* 2, 23–26, and Wilkinson, *Orientalism, passim.*
53. Tasfā Ṣeyon, unpaginated Ge'ez introduction; translation in Lefevre, "Notizie," 90.
54. Wijnman, *Books on the Orient,* xix; Hamilton, "Eastern Churches," 237–239.
55. In 1518 and 1522 he apparently used the type to publish revised editions of the Psalter in Germany: Wijnman, *Books on the Orient,* xv.
56. Hiob Ludolf, *A New History of Ethiopia, being a full and accurate description of the kingdom of Abessinia, vulgarly, though erroneously call the Empire of Prester John,* trans. J.P. Gent (London: Samuel Smith Bookseller, 1682).
57. Ṣagā Za'āb, 546–548.
58. Some of these earlier testimonies are described in Salvadore, "Ethiopian Age of Exploration."
59. Ludolf, *New History,* 382, 78, 82–88.
60. Ibid., 193, 220–226, 225, 307. The origin of these references is not always clear: they could be derived from the oral testimony of Gorgoryos, the manuscripts of Santo Stefano, or still other sources.
61. Ludolf, *New History,* 383, 257.
62. Ibid., 75, 78–79.
63. Ṣagā Za'āb, 577–579.
64. Ibid., 550.
65. Lefevre, "Riflessi," 264.
66. On these currents in Renaissance orientalism, see Irwin, *Dangerous Knowledge,* 54–85, and Wilkinson, *Orientalism, passim.*

67. Michael Geddes, *The Church History of Ethiopia* (London: 1696), unpaginated introduction; Hamilton, *William Bedwell*, 69–80.
68. For Irwin, the first true orientalist is Guillaume Postel: Irwin, *Dangerous Knowledge*, 66.
69. Sanjay Subrahmanyam, "Connected Histories: Notes towards a Reconfiguration of Early Modern Eurasia," *Modern Asian Studies* 31, no. 3 (1997): 735–762. For a comparative overview, see Felipe Fernández-Armesto, *Pathfinders: A Global History of Exploration* (New York: W.W. Norton, 2007).
70. Overviews of these developments from a European perspective include Anthony Pagden, *The Fall of Natural Man: The American Indian and the Origins of Comparative Ethnology* (Cambridge: Cambridge University, 1986), and Joan Pau Rubiés, *Travel and Ethnology in the Renaissance: South India through European Eyes, 1250–1625* (Cambridge: Cambridge University, 2000). On the contemporaneous interest in travel writing in Asia, see Muzaffar Alam and Sanjay Subrahmanyam, *Indo-Persian Travels in the Age of Discoveries, 1400–1800* (New York: Cambridge University, 2007) and Nabil Matar, ed. *In the Lands of the Christians: Arabic Travel Writing in the Seventeenth Century* (New York: Routledge, 2003).
71. Against an earlier view that European exploration rapidly destabilized existing explanations for human difference and diversity, Michael Ryan argues instead that awareness of hitherto unknown people did not immediately transform European values and beliefs—instead, scholars began to focus intently on assimilating new information into existing genealogical frameworks: Michael Ryan, "Assimilating New Worlds in the Sixteenth and Seventeenth Centuries," *Comparative Studies in Society and History* 23, no. 4 (1981): 519–538. In a similar vein, Anthony Grafton suggests that exploration led many European scholars to the project of fitting "new masses of difficult data to the inherited shapes of learning," and that this eventually led to new approaches to the classical tradition. See Anthony Grafton, *New Worlds, Ancient Texts: The Power of Tradition and the Shock of Discovery* (Cambridge, MA: Harvard University, 1992), 93.
72. Hirsch, *Damião de Gois*, 74.
73. By 1543 the library of the Jesuit College in Goa possessed a copy of Ṣagā Za'āb's work. Pennec, *Des jésuites au royaume du Prêtre Jean (Éthiopie)*, 32–35.
74. Boemys, *Manners*.
75. Cf. Morton Eudes Geddes, *Catholique traditions, Or A treatise of the beliefe of the Christians of Asia, Europa, and Africa, in the principall controversies of our time In favour of the lovers of the catholicke trueth, and the peace of the Church* (1609), and Edward Brerewood, *Enquiries touching the diversity of languages, and religions through the cheife parts of the world* (London: 1614); Ludolf, *New History*, 238–239.
76. Irwin, *Dangerous Knowledge*, 246; Lefevre, "Notizie," 107–108.
77. Brian Walton, *Biblia sacra polyglotta: complectentia textus originales, Hebraicum, cum Pentateucho Samaritano, Chaldaicum, Graecum. Versionumque antiquarum, Samaritanae, Graecae LXXII Interp., Chaldaicae,*

Syriacae, Arabicae, Æthiopicae, Persicae, Vulg. Lat. quicquid comparari poterat (London: Thomas Roycroft, 1657). See also the discussion in Wijnman, *Books on the Orient,* xxv.

78. Mariano Vittorio, *Zentu maṢehafe temeheret zalsana ge'ez zāysemay kaledawi hadis ser'at tagabra kem yetemaharā 'ele 'iye'amerā senāy we'etu tegebre ka'ey māriano witori'o zara'etu. Chaldeae seu Aethiopicae linguae institutions. Opus utile, ac eruditum* (Rome: Typis Sac. Congregationis de Propaganda Fide, 1630). See also the discussion in Guidi, "La prima stampa del Nuovo Testamento in etiopico," 274, n. 2.
79. For a summary of these later editions, see Wijnman, *Books on the Orient,* xxvi.
80. For the negative assessments of Nicolao Godinho and Jeremiah Lobo, see *AE,* 2:822, 855–856, respectively.
81. Ṣagā Za'āb, 576; Ludolf, *New History,* 238–239. Details on Matéwos can be found in Richard Pankhurst, "The History of Ethiopian-Armenian Relations (I)," *Revue des études arméniennes* 13 (1978/1979): 279.
82. Potken, unpaginated introduction, and Teseo Ambrogio, *Introductio in Chaldaicam linguam, Syriacam, atquen Armenicam, et decem alias linguas* (Pavia: 1539). See also Robert Wakefield, *Sacrarum literarum professoris eximij oratio de laudibus & vtilitate trium linguarum Arabicae Chaldaicae & Hebraicae atque idiomatibus hebraicis quae in vtroque testamento inueniuntur.* (London: Apud Vinandum de Vorde, 1528).
83. De Góis, *Fides,* 52 and Vittorio, *Zentu maṢehafe.*
84. Ludolf, *New History,* 75.
85. Ṣagā Za'āb, 576; Ludolf, *New History,* 152–153; an example is Edward Brerewood, *Enquiries touching the diversity of languages, and religions through the cheife parts of the world* (London: 1614), 163. Like "Chaldaean," the idea of Prester John rested upon medieval confusions about the distinction between Ethiopia and India: Donald Lach, *Asia in the Making of Europe* (Chicago: University of Chicago, 1994), 2: Bk. 3: 514–515.
86. Edmund Bohun, *A geographical dictionary representing the present and ancient names of all the counties, provinces, remarkable cities, universities, ports, towns, mountains, seas, streights, fountains, and rivers of the whole world* (1693), 4.
87. Grafton, *New Worlds, Ancient Texts,* 106.
88. Even in 1634 a century after the first Ge'ez publications in Europe, the Vatican was unable to find an interpreter for visiting Ethiopian dignitaries: Lefevre, "Riflessi," 275.
89. Ludolf, *New History,* 242.
90. For examples, see *Testamentum Novum,* 131–132, which contain images of the angels Gabriel (Gabre'él), Michael (Mikā'él), and Raphael (Rāfā'él).
91. Ibid., 63–64.
92. Ibid., 62–63.
93. Ibid., 64.
94. Sevir Chernetsov, "Ethiopian Theological Response to European Missionary Proselytizing in the 17th-19th Centuries," in *Ethiopia and the Missions,* 53–62. See also Pennec, *Des jésuites au royaume du Prêtre Jean (Éthiopie),*

298, and Merid Wolde Aregay, "The Legacy of Jesuit Missionary Activities in Ethiopia from 1555–1652," in Getatchew Haile, Aasulv Lande, and Samuel Rubenson, eds. *The Missionary Factor in Ethiopia: Papers from a Symposium on the Impact of European Missions on Ethiopian Society, Lund University, August 1996* (New York: Peter Lang, 1998), 41.

95. Pankhurst, "The History of Ethiopian-Armenian Relations (I)," 292.
96. Edward Ullendorf, "The Confessio Fidei of King Claudius of Ethiopia," *Journal of Semitic Studies* 32, no. 1 (1987): 159–176.
97. Ṣagā Za'āb, 578.
98. In some sense we can say that he accomplished his goal—the Catholic Church recognized his text as an authentic and thus heretical statement.
99. Guidi, "La prima stampa del Nuovo Testamento in etiopico," 273.
100. Tasfā Ṣeyon, 133.
101. *Testamentum Novum,* 7–12 . On the Ethiopian use of the canon tables, see *AE,* 1:680, and 2:455.
102. One could argue that the inclusion of a Ge'ez grammar suggests a European audience for the work. While this is certainly plausible, it is also true that an Ethiopian reader could have found the prefatory Ge'ez grammar as useful as a European would, given that Ge'ez was scholarly language and not a vernacular. Indeed, in this same period Ludolf noted that Ge'ez was known as "the language of books; either because it is only us'd in Writing, or else because it is not to be attained without Study and Reading of Books." Ludolf, *New History,* 73.
103. See especially C.A. Bayly, "Archaic and Modern Globalization in the Eurasian and African Arena, ca. 1750–1850," in Anthony G. Hopkins, ed. *Globalization in World History* (London: Norton, 2002), 45–72; and Anthony G. Hopkins, "Introduction: Interactions between the Universal and the Local," in Anthony G. Hopkins, ed. *Global History: Interactions between the Universal and the Local* (New York: Palgrave Macmillan, 2006), 1–38. In addition to these collections, see also Geoffrey Gunn, *First Globalization: The Eurasian Exchange, 1500–1800* (Oxford: Rowman Littlefield, 2003), Luke Clossey, *Salvation and Globalization in the Early Jesuit Missions* (Oxford: Oxford University, 2008), and *Journal of Global History* 2, no. 2 (2007), special issue on "Islamic History as Global History."
104. I borrow this term from Clossey, *Salvation and Globalization in the Early Jesuit Missions,* 249–256.
105. On the connections between messianism and European exploration, see Sanjay Subrahmanyam, *The Career and Legend of Vasco da Gama* (New York: Cambridge University, 1997), 24–75.
106. Wilkinson, *Orientalism,* 189–192; *AE,* 3:601–603.
107. Murad Kamil, "Translations from Arabic into Ethiopian Literature," *Bulletin de la société d'archéologie Copte* 7 (1941): 61–71.
108. Gunn, *First Globalization,* 1–8.
109. Ṣagā Za'āb, 567–568.

CHAPTER 9

"In manners they be rude, and monst'rous eke in fashion": Images of Otherness in Early Modern Drama

Patrick Tuite

From the end of the sixteenth to the beginning of the seventeenth centuries, England challenged Spain for control of the seas, explored Africa and the New World, and extended its power across the British Isles to become an empire with a growing number of colonies. As the British Empire expanded, its subjects encountered a variety of different races and cultures. Pamphlets, histories, maps, and plays attempted to define the characteristics of these aliens, describe their manners and customs, and organize them within a hierarchy based on the social, religious, and political values that ordered Great Britain's subjects. In his *Britannia*, one of the most popular histories written during this period, William Camden claims that the British were already a nation that combined the best traits of the Britons, Romans, Saxons, Danes, and Normans.

According to Camden, Great Britain's continued prominence depended upon the ability of its subjects to maintain their laws, culture, and religion while also translating or incorporating people from other nations who could conform to them. Camden's terms name different but related processes: translation describes how the descendents of immigrants from a select number of nations could become British through acculturation and intermarriage, while incorporation explains how Great Britain's leaders allowed foreigners to maintain aspects of their native culture if they could serve the empire and obey its laws. To use Camden's logic, if aliens could not become British or live peacefully among them, they should be displaced or suffer violent subjugation. The task was to discern who could or could not be included as a subject within the expanding British Empire.

This chapter examines how a select number of theatrical performances staged in early modern London imagined the world and its different inhabitants in relation to an evolving Anglo-Protestant social order. Many of the plays that feature strangers from distant lands portray them as either capable of being incorporated into British society or essentially alien and unable to successfully assimilate into the same. Images of these foreigners were available to literate British subjects through a variety of publications. However, while woodcuts and engravings from this period depict natives from exotic locations with vivid detail, actors in London's public theaters animated the same characters, giving them an amusing or alarming immediacy through their performance on stage. Given the limitations of the early modern theater to depict specific locations, it is important to explain how theatrical performances indicated racial and cultural difference during this period and how these characterizations reflected changing notions of British subjectivity.[1] By identifying and analyzing the Irish characters that appear in plays staged in London between 1596 and 1614, it is possible to explain how the city's theater companies appropriated certain stereotypes found in popular publications to construct Ireland's different inhabitants as either comical servants who deserved a subservient position within the empire's dominions or frightening enemies who threatened to undermine the authority of its British leaders.

Performing Blackness, Savagery, and Civility

Two plays staged before King James in 1604 and 1605 help explain how theatrical performances used popular stereotypes to define different aliens as capable of becoming civil subjects or ultimately irreformable. The traits that identified these characters as docile or dangerous highlight similar markers that shaped the performance of different Irish identities in early modern London. Public performances and courtly entertainments staged at the beginning of King James's reign deployed changing notions of blackness to display the grandeur of his empire and the superiority of his conforming Protestant subjects. Those spectacles that featured African characters employed changing definitions of white and black and the shifting spaces that fair and black peoples occupied in the early modern imagination. Until the mid-sixteenth century, a person described as fair was considered beautiful, and the term applied to people of different races. Labeling people as black or dark indicated their lack of civility. After the beginning of the English slave trade in Africa, fair identified a person as light-skinned, and by the early seventeenth century, the same term was limited to only white, Christian Europeans.[2]

Shakespeare wrote *Othello* when the English monarchy was actively promoting colonial projects but had not fully invested in the African slave trade.[3] The play reflects these developments and presents a complex imperial discourse in which race, culture, and religion define which characters appear to be fair or dark, civil or savage. In the play, Venice's leaders appoint Othello as the military governor of Cyprus. The Doge and his senators recognize that the Moor is an outsider, but they argue that he is a valiant soldier who has faithfully served the Venetian Republic. Though Othello was raised in North Africa and sold into slavery, he is of noble birth. When the Venetians released him from his bondage and welcomed him into their ranks, he was able to civilize his speech and manners. Othello expresses his gratitude for the kindness of his masters and describes that he has served them by fighting cannibals, anthropophagi, and other strange natives in rocky hills, mountain caves, and empty deserts to further expand Venice's empire.[4] Othello's acculturation and exploits in arms distance him from the savage peoples he defeated along the margins of the civilized world and strengthen the bonds that unite him with his Venetian masters.

Yet, as the governor of Cyprus, Othello must lead a combined force of Venetian and Cypriot soldiers against an impending Turkish invasion. In essence, the Doge and his senators ask a Moor to defend a Christian colony from an invading Islamic army. Because of his race and the religion of his ancestors, Othello appears to be more like the Turks than his Venetian masters. What then could stop this stranger from reverting to his native culture and religion and betraying his subordinates? Othello does not have to fight the Turks, but he proves to be a tragically flawed leader. His noble ancestry, the climate of his native land, and the harsh experiences of his upbringing make him a perfect soldier. However, the same traits undermine his ability to maintain a civil and Christian family, and although Shakespeare's play portrays him as a loyal surrogate of the Venetian Republic, it ultimately reinforces the divisions separating white Europeans from nonwhite people living outside of the continent.[5]

Iago, Roderigo, and Brabantio argue that Othello should never have assumed the authority that the Venetians gave him. According to their logic, Othello's downfall is a consequence of improper colonial positioning, and his appointment as the military governor of Cyprus reflects a misunderstanding of the inherent inferiority of nonwhite, non-European people. If Othello had remained a warrior fighting heathens along the margins of the empire, a more appropriate Venetian aristocrat would have married Desdemona and governed Cyprus. By welcoming Othello into their society, appointing him as a military governor, and allowing him to marry the daughter of a noble senator, the leaders of the republic put Cyprus at risk and hastened Desdemona's doom. A conservative reading of the play's political lessons indicts Venice's leaders, including the Doge, Brabantio, and Cassio, for not better managing their subordinates. Iago argues that Desdemona's marriage to the Moor is a "gross revolt" and prophesizes that their unnatural union will produce uncivil heirs.[6] When Brabantio learns that Desdemona has secretly married the Moor, he rues his lack of vigilance and weighs the political and social implications of his daughter's blind obedience to a wayward stranger:

> Mine's not an idle cause. The Duke himself,
> Or any of my brothers of the state,
> Cannot but feel this wrong as 'twere their own;
> For if such actions may have passage free,
> Bondslaves and pagans shall our statesmen be.

Administrators in Elizabethan and Jacobean England used similarly conservative arguments to prevent people of different cultures and religions from being appointed into positions of power. This was especially true in Ireland, where Catholics and dissenting Protestants were barred from military commissions and important government offices. In 1676, Thomas Rhymer argued that Shakespeare had marred Cinthio's play by offering Othello a title and station, "which neither history nor heraldry can allow him."[7] According to Rhymer, the English would never have appointed a "Blackamoor" to lead an army or allow him to marry the daughter of an aristocratic leader. According to this argument, people of different races could serve the English monarchy as subordinates, especially abroad. Yet, when similar policies offered proxies a limited position within Great Britain's evolving social structure, its leaders imagined these surrogates as inferior. Thus Shakespeare's *Othello*, though it presents the Moor as a complex and sympathetic character, ultimately excludes him from the highest ranks of the Venetian republic. In other words, soldiers like Othello could serve the British Empire but could not be fully translated into British subjects.

Staged before James I at Whitehall just two months after the first performance of *Othello* at the Globe, *The Masque of Blackness* also rehearsed competing theories of blackness, but it contended that the power of the Jacobean court could translate noble black natives into civil subjects.[8] In Jonson's masque, the daughters of the river Niger learn of their inferiority through a waking vision, in which they are instructed to find an island-nation whose name ends in "Tania." The young women travel from Ethiopia, the darkest nation in the world, in search of the fair land. Having no success in Mauritania, Lusitania, and Aquitania, lands in which the inhabitants become lighter by degrees, Oceanus informs Niger that Albion is the island that his daughters seek.[9] When they reach the fair land, radiant beams emanating from its monarch promise to "salve the rude defects" of the Ethiopian nymphs and blanch their skin.[10]

In *Othello* and the *Masque of Blackness* two aliens attempt to marry into a white, Christian, and European society, one literally and the other symbolically. Both are Africans of noble ancestry who attempt to assimilate into a white society, but only one can fully integrate into it. The careful arrangement of race, gender, and civility in each union makes one marriage natural and the other unnatural. Shakespeare's play ends in tragedy as a consequence of the Moor's innate inferiority and subjugated position within Venetian society. In Jonson's masque, God's anointed monarch leads Christian Europe, and his civility allows him to translate a dark pagan from Africa into a conforming white subject. The masque imagines the British Empire as a growing family that is led by an appropriately appointed patriarch. Elizabethan and Jacobean authors developed similar tropes to describe the relationship between the English monarchy and Ireland's inhabitants.

While the Irish were the second (after the Welsh) to bear the brunt of English colonialism, their portrayal in a variety of English cultural productions set the stage for later encounters. Depictions of the Irish as savage, bestial, and uncivilized provided the groundwork for later descriptions of other apparently inferior races. These anthropological accounts emphasize the similarities that the English believed they had found between Ireland's seemingly white and Christian natives who violently resisted British colonization and the black Africans the English sold into slavery. This textual magnification of radical racialized alterity supported further violence, reinforcing claims that the uncivilized natives of some regions could be saved from their inherent inferiority through contact with the great English sovereign while others were so savage that their only hope for salvation was complete domination by the English crown. The Irish, despite their geographic proximity, white skin, and Christian traditions, fell into the latter camp, while some Africans, Asians, and Middle Easterners, especially those less resistant to British colonial projects, were deemed capable of reform.

Identifying the Civil and Wild Irishman

Judgments concerning the relative civility or savagery of different races reinforced British imperial projects. The image concepts that determined these judgments were based upon a combination of biblical exegesis, historical accounts of past encounters with different races and cultures, and anthropological descriptions of these aliens. Such texts include descriptions of the natives who inhabited the wilder parts of the British Isles, and near the end of the sixteenth-century histories concerning Ireland were among the most popular of these texts. Anglo-Norman lords had invaded Ireland centuries before the English explored Africa or America, and while attempting to conquer the island, they described Ireland's natives in the same language that British colonists used to label Africans and Indians.

Giraldus Cambrensis was the first British historian to describe Ireland as a distinct island-nation. Giraldus worked in the court of Henry II and his family was directly involved in the invasion of Ireland. Norman lords living in Wales had begun the campaign before King Henry ratified their plans, and Giraldus' firsthand account justifies the subjugation of Ireland's Christian inhabitants by making their island appear to be the secret lair of freaks and savages.[11] According to Giraldus, before the Anglo-Normans invaded Ireland, the Irish had no culture and lived like beasts. They wore no mantles or cloaks and rode without saddles, spurs, or leggings. They went into battle naked and were normally "made up in a barbarous fashion."[12] The barbaric state of the Irish was a result of their isolation, and this condition prevented them from developing like other European peoples.[13] Writers in early modern England referred to Giraldus's work when they described Ireland's history. His dire assessment of the Irish was especially useful when Tudor and Stuart monarchs attempted to pacify and reform Ireland's inhabitants. Like Giraldus, Elizabethan and Jacobean authors claimed that the Irish were a distinct and barbarous race in need of reform. Between 1596 and 1614, a variety of histories, travelogues, maps, poems, and plays featured Giraldus's description of the Irish. Manuscripts, printed texts, and theatrical performances helped these accounts circulate through London and made them available to a wide demographic.

Throughout the period covered in this chapter, British historians labeled Ireland's three communities as Old Irish, Old English, and New English. According to these accounts, race, culture, and religion separated one group from another. Moreover, an astute observer could discern the differences among the kingdom's different inhabitants by noting their language, fashion, manners, and customs.[14] Ireland's natives constituted the Old Irish community. They were Gaelic by culture and Catholic by religion, and they constituted the majority of Ireland's population. The Old English were the descendents of the Norman conquerors that Giraldus had valorized. Some of these former settlers were Anglo by culture, but many lived among the Old Irish and adopted Gaelic names, language, fashion, and customs.[15] Many were also Catholic. As a result, New English settlers considered them as neither English nor Irish, but a degenerate and dangerous mixture of "English-Irish."[16]

Near the end of the sixteenth century, the concern among Elizabethan undertakers and administrators in Ireland was whether or not they could reform the kingdom's uncivil Old Irish and Old English subjects. Among the British who worked in Ireland, one faction argued that a majority of the Irish were willing to learn English, desired peace, and accepted the authority of English law.[17] This faction claimed that popular accounts of Ireland misrepresented its inhabitants; it was the miserable condition of Ireland's poor subjects and not their nature that had driven them to rebellion in the past.[18] Even when Hugh O'Neill, the earl of Tyrone, rebelled at the end of the century, Captain Thomas Lee appealed to Queen Elizabeth that if her agents reformed their policies, the chieftain and his allies would embrace the Queen's laws and live peacefully among Ireland's civil subjects.[19] According to Lee, there were only two options for Ireland's rebellious subjects. He wrote to the Queen that they must be "utterly overthrown or cut off, or else drawn to subjection and due obedience to your majesty and your laws."[20]

O'Neill did not remain loyal to Queen Elizabeth, and his defeat in 1603 discredited the attempts of Lee and like-minded administrators to employ Old Irish chieftains as surrogates. Lee's detractors believed that the Old Irish would remain uncivil unless they were first subdued and then taught to obey English laws.[21] Accounts detailing the failure of the Old English to conquer Ireland's natives in the past supported more militant colonial policies, and it is important to compare the political agenda promoted in Jonson's masque to the warnings concerning Ireland's subjects found in Tudor histories. In the masque the power of the British monarchy is able to blanch the defects of an Ethiopian princess and make her white. Near the end of the sixteenth century, English historians portrayed Ireland's Old English settlers as becoming less civilized after they traveled to Ireland and lived among its savage natives. While Jonson's masque presents the transformation of the Ethiopian process as forward progress, English historians claimed that Ireland's climate and the brutishness of its inhabitants could make the English more like the Irish. Writing from Munster near the end of the sixteenth century, Edmund Spenser warned that living among the island's backward natives could alter the nature of even the most civil subjects.[22]

Two images display the results of this forward progress and degeneration. In the first, Queen Anne of Denmark appears in the costume and makeup that she wore while playing the daughter of Niger in Jonson's masque. In the second, Thomas Lee appears as the Captain General of the kern in Ireland. The print and the painting capture rare celebrations of cultural hybridity, but Lee's portrait imagines a more complex transformation. In it, the captain stands poised between two worlds. A dark wood and standing pool looms above him and stretches to his left, while the sun shines on open fields and distant mountains to his right. Marcus Gheeraert created the portrait at Lee's request in 1594, and its composition is similar to Gheeraerts' Ditchley portrait of Queen Elizabeth. In each, a confident leader guards a peaceful landscape against a gathering darkness. Lee's peaceful landscape is, in fact, the Ireland inhabited by conforming subjects. He stands tall, like the oak tree immediately behind him, armed with a spear and pistol to protect Ireland's loyal subjects from the rebels and thieves that British historians claimed infested the island's woods, hills, and bogs.[23] Despite the danger that this setting suggests, Lee does not wear the armor or carry the weapons of a typical Elizabethan adventurer. Like an actor in one of Jonson's masques, he appears to have stepped out of the woods costumed in the habit of an Irish kern, with his chest and legs exposed and his loins just barely covered by the tails of his shirt. The captain's casual pose and the whiteness of his bare skin contrast sharply with the painting's dark and ominous woods. Although Lee employed a court artist to paint his portrait, the creation of his hybrid character offers important insights into the representation of foreign racial groups in theatrical performances staged in London during the same period. Viewed after the captain's humiliating

Figure 9.1 Anne of Denmark as the Ethiopian Princess in Jonson's *Masque of Blackness*, in a rendering by Inigo Jones, 1605, by permission of the Folger Shakespeare Library.

downfall, the painting suggests that instead of making his Irish subordinates English, Lee's prolonged exposure to the kingdom's natives had made him Irish. According to more conservative writers, this unraveling was a serious concern as other English settlers had degenerated into the worst of Ireland's rogues in the recent past.[24]

According to Camden, the kingdom's "Wild Irish" consisted of the rebellious Old Irish and their Old English allies.[25] They lived beyond the English Pale and fiercely guarded their Brehon laws, Gaelic culture, and superstitious religion.[26] Despite the efforts of the kingdom's Protestant authorities to convert them, a majority of the Old Irish and Old English nobles remained Catholic, and they maintained close ties to Catholic institutions in Europe. Whenever a European power threatened England, Ireland's Protestant administrators regarded these lords with suspicion. The degenerate Old English were especially dangerous. They had formed strong alliances with the members of Ireland's other communities by marrying into the most powerful Old Irish and New English families. This pragmatism allowed many of the Old English to prosper under different English administrators and then oppose them when circumstances suited their interests. At the end of the sixteenth century, Ireland's Protestant subjects believed that the Old English were among the most notorious of the kingdom's rebels.[27] Some had degenerated to the point that they could not be distinguished from the kingdom's uncivilized Old Irish natives.[28] Yet other Old English lords were loyal to the Tudor and Stuart monarchs, and they secured their wealth, status, and privilege by cultivating the language, fashion, and manners of Ireland's British authorities.

The New English included those English, Scottish, and Welsh Protestants who came to Ireland during or after the reign of the Tudor monarchs. Seventeenth-century religious differences split the New English into conforming and dissenting Protestants. Conforming Protestants adhered to the practices of the Church of Ireland, and they occupied the most powerful positions in the kingdom's carefully guarded institutions. However, dissenting Protestants, from Scottish Presbyterians to English Quakers, eventually earned limited enfranchisement. New English and conforming Old Irish and Old English lords implemented colonial policies in Ireland. Though the rebel Irish robbed and murdered innocent settlers, the New English justified atrocities committed against the Old Irish and Old English rebels by highlighting their brutish natures and obstinate allegiance to Catholicism and the tyrannical foreign powers that maintained its sway over most of the kingdom's subjects. As a result of their efforts to reform the kingdom for their Protestant monarchs, the New English frequently reaped great benefits in terms of seized land, titles, and wealth. They were not, however, the sole recipients of the monarchy's largesse, and the inconsistent and shifting nature of the social and religious hierarchies that attempted to order Ireland during the seventeenth century are worth noting.

Ireland's administrators used the characteristics associated with the Old Irish, Old English, and New English to identify and position each subject within the kingdom's social structure. From the Reformation to roughly the end of the eighteenth century, Ireland's rulers ranked the island's three communities in the following order: New English, Old English, and Old Irish. New English aristocrats had the greatest power, but certain factors complicated this hierarchy. For most of the seventeenth century, conforming Protestants had more political power than dissenting Protestants in both England and Ireland. In addition, a select number of Old Irish and Old English lords retained seats in the Irish Parliament. Some converted to the Church of Ireland, and conforming Anglo-Protestant leaders offered these converts landed titles and political enfranchisement, while barring many Catholics and dissenting Protestants from the same privileges. Changing historical circumstances further complicated how Ireland's leaders ranked their subjects, and the relative power of each community rose and fell with every conflict and regime change that occurred in Ireland and Great Britain during this period.

In 1611, John Speed published a collection of maps aptly entitled the *Theatre of the Empire of Great Britain*. The relationship between Speed's maps, contemporary histories of Ireland, and the plays that depicted the Irish during this period cannot be overstated as the products of each medium provided popular representations of the kingdom's inhabitants.[29] The frontispiece of Speed's *Theatre of the Empire of Great Britain* displays a triumphal arch in which a Britain, Roman, Saxon, Dane, and Norman appear in alcoves. The image replicates the dramatic display of power found in a royal entry and similar processions that celebrated the power of the English monarchy. In this respect, Speed positions the viewer as a privileged participant in a virtual reproduction of the event. Like a monarch traveling across the kingdom in a royal progress, the viewer passes through the triumphal arch and observes each region of the empire, noting the manners and habits of the inhabitants as they greet the viewer. What is important to note is that the characters standing in Speed's first arch represent Great Britain's historic ancestors. Like Camden, Speed claims that the amalgamation of these different peoples made Great Britain the most powerful empire in the world.[30]

Speed did not advocate for the inclusion of a distinctly Irish subculture into Great Britain's dominant social order. He argued that the native Irish were a superstitious and idolatrous people whose ancestors oppressed the island with "rapine and violence."[31] Speed's maps include images of Ireland's natives in which they more closely resemble Indians from Florida and Mexico than the kingdom's English overlords. The engravings attempt to illustrate the similarities among uncivilized peoples, despite the color of their skin.[32] They also highlight the differences between truly civilized white people, notably Anglo-Protestants, and the outwardly fair but innately dark, "Wild Irish." These apparently authentic engravings of Ireland's different subjects graphically illustrate the social structure that English administrators attempted to impose upon the kingdom. Speed inserted a cartouche into the bottom left corner of his map of Ireland. The cartouche has two columns, each of which is subdivided into three frames. A figure appears in each of the six frames with a man and woman representing the kingdom's New English, Old English, and Old Irish subjects. A "Gentleman of Ireland" wearing a fashionable hat, fringed cloak, collared shirt, and fine shoes appears at the top, while a "Wilde Irishman" with unkempt hair, wool mantle, and bare feet is displayed at the bottom.[33]

Speed's taxonomy represents Ireland's early modern social structure as natural and stable, but it was not. In fact, contemporary historians revealed that subjects could move up or down Speed's imaginary scale, and this indeterminacy alarmed English administrators and historians. While a select number of Ireland's inhabitants could be reformed, English settlers living among them could degenerate, becoming more barbarous than the island's brutish natives. Jonson's *Masque of Blackness* dramatizes how this dynamic could in Camden' terms translate the island's darker yet noble inhabitants into conforming and loyal subjects. However, instead of relying on the power of King James to transform them, English officials, historians, and playwrights claimed that if English laws could be fully enforced throughout Ireland, they could subdue its worst natives and redeem its more compliant subjects. The administrators responsible for Ireland's government promoted roughly two sets of policies to achieve this reformation.

A more conservative group attempted to strengthen the boundaries pictured in Speed's map. These officials argued that firm laws should separate Ireland's civil subjects and protect them from the degenerate Old English and backward Old Irish. More progressive administrators claimed that, with proper resources and management, a select number of subjects living in Ireland could become more like the English. In this regard, Speed's engravings dramatically illustrate the conservative approach to Ireland's government and Lee's portrait represents a more progressive and inclusive set of policies. Speed's cartouche replaces mobility with stasis, making any movement among the members of Ireland's three communities appear dangerous, while Lee's portrait boldly crosses these boundaries to display the positive aspects of hybridity.

An anonymous play staged in London in 1596 gives these boundaries even greater emotional force. *The Famous History of the Life and Death of Captain Thomas Stukeley* chronicles the actions of a brave but wayward young gentleman who squanders his father's fortune, marries the daughter of a London alderman, and then leaves his young bride in England to fight in the Irish wars.[34] After defeating Shane O'Neale outside Dundalk, Stukeley leaves for Spain and eventually leads a mixed force of Italians

Figure 9.2 John Speed, *Theatre of the Empire of Great Britain*, 1611, by permission of the Folger Shakespeare Library.

and Germans in a desperate battle against an army of Moors and Turks in Africa. The setting, characters, and actions of this play are similar to those found in Shakespeare's *Othello*. In each play noble characters fight against barbaric enemies but are also responsible for the early deaths of their wives. The difference between the two officers is that while the Moor cannot overcome the physical differences that separate him from the Venetians, Stukeley is an Englishman who reforms his manners and eventually restores his reputation. In terms of cultural boundaries, Othello commands Cyprus from within its walls, but he is still an outsider in Venetian society. On the contrary, Stukeley begins the play at the center of the English empire, a gentleman of means studying at the Innes of Court, but his uncivil actions drive him from his wife and England.

Stukeley uses his wife's wealth to arm a company and join an expedition bound for Ireland. Once he arrives in the troubled kingdom, he challenges the commander of the garrison at Dundalk to a duel. Stukeley's treasonable offense forces the commander to remove him from the city, and Stukeley must defend himself against the Irish outside Dundalk's walls. Therefore, *Othello* and *Captain Thomas Stukeley* are siege dramas in which savages threaten to overcome the defenses of civilized settlers. However, Stukeley's actions at Dundalk place him outside of the law. His insubordination within the garrison makes him as dangerous as the Irish rebels that surround the city, and he must fend for himself outside Dundalk's walls. His reputation lost, Stukeley leaves his company in Ireland the day after he lands at Dundalk, and he does not reclaim his virtue until he fights Moors and Turks in Africa on behalf of the king of Portugal.

Though *Captain Thomas Stukeley* includes a Moorish king who is a loyal vassal of the Portuguese, it depicts other Irish, Moorish, and Turkish characters as bloodthirsty enemies of Christian Europe. In the second act, Shane O'Neale and his rebellious Old Irish chieftains skulk outside of Dundalk at night, hoping that their spies will direct them to a secret entrance. They plan to surprise the English within the garrison and cut off their heads before they muster. A vigilant guard prevents the massacre, and O'Neale is forced to attack the city with kerns and gallowglass from Ulster, using a bagpipe and drums to signal the charge. At Alcazar, Muly-Mahamet proves to be as villainous as O'Neale. In the final acts of the play, the Moor claims that he will support Sebastian, the king of Portugal, by deposing his brother, the Moorish king, and reducing his subjects to the will of the Portuguese. However, Muly-Mahamet's brother is a virtuous leader and friend of the Portuguese and their Spanish masters. Muly-Mahamet deceives Sebastian and urges him to assemble an army and invade his brother's kingdom. When the Portuguese land on the Barbary Coast, they discover that the Spanish will not support the expedition. Muly-Mahamet abandons Sebastian outside Alcazar, and an army of Moors and Turks overwhelms the hapless Christians.

Stukeley leads the foremost company of Christians at Alcazar, and meets a friend from London after the battle. The English gentleman, Vernon, had vied for the affections of Stukeley's wife, but introduced her to him when she confessed that she loved the valiant Stukeley. After Stukeley abandoned young Nell, Vernon vowed to never associate with the rash spendthrift again. Alone on the plains outside Alcazar, Stukeley repents the actions that drove him from his wife and England. The friends reconcile and prepare to die among the "bloody and uncivil Turks." Stukeley's Italian mercenaries kill the Englishmen instead. O'Neale and Muly-Mahamet are also killed by their supposed allies, and their ignominious deaths serve as an example of what should happen to treacherous subjects. After O'Neale is defeated at Dundalk, he approaches the city with a halter around his neck hoping to reconcile himself with the Queen's officers. On the way to Dundalk, he encounters Alexander Mack Surlo, a Scottish lord. Mack Surlo remembers that O'Neale had murdered his cousin, kills the Irish rebel, and sends his head to the English Deputy as a warning to other traitors. After the battle of Alcazar, the loyal Moorish king vows to capture Muly-Mahamet, flay him, and have his body carried throughout the kingdom to display the consequences of such treachery.

Unlike *Captain Thomas Stukeley*, Shakespeare's *Henry V* and Jonson's *Bartholomew Fair* feature Irish characters that faithfully serve their English masters. Though these depictions are comical, the plays supported policies that gave faithful Irish subjects positions that offered them limited authority in Ireland and Great Britain. In addition, the plays were written and first performed at times when Ireland was experiencing rapid social and political change. Though Shakespeare did not have access to Speed's maps when he wrote *Henry V* in 1599, he used Raphael Holinshed's *Chronicles of England, Scotland and Ireland* to describe some of the play's events. By 1578, Holinshed's *Chronicles* included a section written by Edmund Campion. Campion's history relies heavily on Giraldus Cambrensis's account of the kingdom and emphasizes the need to civilize the Old Irish and degenerate Old English. In addition to Campion's section in Holinshed's *Chronicles*, William Camden and John Derricke also published versions of Ireland's history before the end of the sixteenth century.[35] Shakespeare capitalized on the popularity of these texts and the general concern about the latest campaigns in Ireland when he wrote *Henry V*.

It appears most likely that Shakespeare wrote *Henry V* when the Earl of Essex was preparing to lead an army into Ireland, and the excitement surrounding this venture may have inspired Shakespeare's history. Essex left England on March 27, 1599 commanding the largest English force ever assembled to defeat Ireland's rebellious Old Irish and Old English lords. The Lord Chamberlain's Men performed *Henry V* in the same month. According to James Shapiro, Shakespeare wrote about young Harry's successes in France in anticipation that Essex would return from Ireland as a triumphant champion of the English monarchy.[36] However, Shakespeare's play reveals the dangers of a costly and difficult campaign as much as it valorizes Essex and his knights. In this respect, the play supports policies that Essex and his English officers, including Captain Thomas Lee, employed to isolate and defeat Ireland's most notorious rebels. Essex, Lee, and their subordinates allied themselves with those Irish lords who demonstrated their loyalty to the English monarchy, and Shakespeare's play advocates the incorporation of loyal Welsh, Scots, and Irish officers into the British social structure under specific conditions.

Gower, Fluellen, Jamy, and MacMorris are English, Welsh, Scottish, and Irish captains who travel to France and fight for the English monarchy. The orthography found in the Shakespeare's First Folio strongly suggests that the playwright and his company used different dialects and key phrases to identify each officer's nationality. The different dialects make each captain distinct, but they still serve one monarch. However, when the captains meet outside Harfluer, Fluellen ridicules MacMorris, calling the bearded Irishman an ass.[37] He insists that all of the captains but MacMorris share an understanding of Roman military discipline. The slander reiterates similar claims made by sixteenth-century historians: the Irish were a barbaric race because the isolated island had not benefited from Roman rule.[38] When this attack does not incite MacMorris, Fluellen questions why so few of his nation fight in Henrys' army. MacMorris famously replies "What ish my nation?" It is an apt remark as MacMorris is "English-Irish." He is not a Gaelic-speaking native of Ireland, but a loyal Old English officer. However, he is not a disciplined leader, and he betrays his lack of civility when he threatens to cut off Fluellen's head if the Welshman insults him again. In *Captain Thomas Stukeley* the English officer is, like MacMorris, undisciplined and does not observe the rules of war. Yet MacMorris's threat reinforces the stereotypes that figure the Irish as essentially violent, making MacMorris more like O'Neale than Stukeley, as both Irishmen vow that they will decapitate their enemies. However, MacMorris appears to have assimilated to the English language and manners as best as he is able, and he faithfully serves his monarch in a foreign war that does not directly concern Ireland.

Fluellen does not argue against MacMorris's loyalty; he is frustrated by the Irishman's elevated status within the King's army. When Gower orders Fluellen to meet with the Duke of Gloucester in the mines, Fluellen indicates that mining is below his chivalric sense of conduct. When Gower mentions that the King's youngest brother "is altogether directed by an Irishman," in this uncivil activity, the Welsh captain becomes furious. His anger comes less from the questionable ethics of mining than an Irishman having authority over an English nobleman. For Fluellen, misplacing MacMorris could have dangerous consequences. Fluellen calls the Irishman an ass and a puppy dog, animals that can be domesticated but lack intelligence and require a strict master. According to Fluellen's reasoning, conforming Irishmen should be included among the loyal subjects fighting for the English monarchy, but they should serve only as subordinates. Inverting the natural relationship between a master and servant could prove disastrous, and, as an allegory representing Ireland's colonization, such improper positioning and lax discipline encouraged degeneracy, and MacMorris's misdirection of Gloucester in the mines reflects that downward movement.

Despite their differences, King Henry's captains provided a fantasy of unity at a time when Anglo-Protestant leaders feared that rebellion among their subjects could invite foreign invasion that would prove disastrous for the English monarchy and its reformed religion.[39] To create his loyal band, Shakespeare appropriated national stereotypes that appear in Richard Johnson's *Famous History of the Seven Champions of Christendom.* In Johnson's text, St. George boasts that he has killed a dragon that ravaged North Africa and won the daughter of the king of Egypt. He also defends her from the Black Prince of Morocco. Later in Johnson's story, the patron saints of Wales, Scotland, Ireland, and France join St. George, and under the leadership of England's patron saint, they defeat an Islamic army from Turkey.[40] Johnson's story identifies the national character of each western European saint and demonstrates St. George's authority over them. St. George unites the saints to fight dark heathens threatening the borders of civilized Europe. Essex was a knight of the Garter and emphasized the order's devotion to its patron saint. In April of 1599, he celebrated St. George's Day in Dublin by knighting eighty-one of his followers, and he styled his campaign in Ireland as another crusade against barbaric heathens.[41] Leaders of the Church of England supported Essex's campaign by claiming that the English army should "punish devils that were once saints."[42]

Though MacMorris and his fellow captains are not saints, they share a similar relationship to one another and their English leader. In addition, Tudor histories explain that MacMorris and St. Patrick are not actually Irish. They are West-Britons. Though St. Patrick and MacMorris come from Ireland, they have English ancestors and serve an Anglo leader. Giraldus claimed that St. Patrick was "Britannic by birth."[43] In the sixteenth century, English historians appropriated St. Patrick as a proto-Protestant reformer who civilized Ireland's idolatrous natives. Such revisionist hagiography made St. Patrick a symbolic ally in England's struggle against the dark forces that infected Ireland and encroached upon Europe's Christian monarchies. In his *Britannia*, Camden praises St. Patrick's efforts and advocates the mingling of conforming subjects representing different races who supported Great Britain's reformed religion.[44] According to Camden's logic, the four captains in Shakespeare's play are worthy subjects who, with the proper discipline and guidance, could be incorporated into a peaceful and industrious society. However, while some of these characters are able to be translated fully into the kingdom's social structure, others could only serve its leaders among its lower ranks. For example, when King Henry woos Katherine at the end of the play, he dreams "shall not thou and I, between Saint Denis and Saint George, compound a boy," who will have the strength to defeat the Turk.[45] Katherine can help Harry produce a strong heir because she is a civilized European aristocrat. Contemporary histories did not recommend that Britain's leaders establish a similar union with Ireland's less civilized inhabitants, whether they were Old Irish or New English.

In November 1599, the Lord Admiral's Men staged the first part of *Sir John Oldcastle, the Good Lord Cobham* at the Rose. The play was successful enough to be performed by the Admiral's Men again in March 1600.[46] At the end of the sixteenth century, the Admiral's Men were the most important rivals of the Lord Chamberlain's Men.[47] The competition between the companies and a desire to please their different patrons inspired the Admirals Men to stage *Sir John Oldcastle* as a response to the history plays that the Lord Chamberlain's Men had popularized.[48] Scholars argue that the Admiral's Men paid Michael Drayton, Richard Hathway, Anthony Munday, and Robert Wilson to write a play that would embarrass the Lord Chamberlain's Men over the Oldcastle controversy. In 1596, the Lord Chamberlain's Men staged Shakespeare's first part of *Henry IV*, in which Oldcastle appears as a drunken thief who frequents brothels. By 1599, the Oldcastle family had become an important part of Queen Elizabeth's court, and Sir William Brooke, the Lord Cobham, served as her Lord Chamberlain from 1596 to 1597.[49] As a result, Shakespeare was forced to change Oldcastle's name to Falstaff.[50] *Sir John Oldcastle* portrays Lord Cobham as a brave and virtuous Proto-Protestant martyr and replaces his besotted fool with a recusant priest named Sir John of Wrotham.[51]

Munday and Lord Cobham appear to have had Puritan sympathies, and as James Marino states, "by reforming Oldcastle, in every sense of the verb, transfiguring him from low comedy to pious historical tragedy, the Admiral's Men simultaneously appealed to the taste for Protestant patriotism and assumed the greater weight of tragedy."[52] Marino's assessment is correct and most of the scholarship concerning the significance *Sir John Oldcastle* compares its main character to Shakespeare's Falstaff, and such analysis identifies how the play signals a shift toward more strident anti-Catholic sentiment in Elizabethan England. However, these assessments often disregard how the events surrounding Essex's Irish campaign shaped the imperial model found in *Henry V* and *Sir John Oldcastle*. According to this reading, *Sir John Oldcastle* is as much a response to the pro-Essex propaganda found in Shakespeare's *Henry V*, as it is a Protestant hagiography that corrects the errors in the first part of *Henry IV*.

The ideological shift from an inclusive to exclusive image of empire is most accessible through the depiction of the Irish in *Henry V* and *Sir John*

Oldcastle. The plays promote competing colonial models that position the Irish in different ways. In Shakespeare's play, an Old English captain from Ireland faithfully serves an English monarch and has the same status as his fellow English, Scots, and Welsh captains, and they battle the enemies of King Henry in France. In *Sir John Oldcastle*, the son of Sir Richard Lee returns from the Irish wars with his servant, Mac Chane. While traveling across Wales, Mac Chane kills his master, robs him of this gold, and leaves his body in a thicket. The dialogue and stage directions included in the first printing of the play indicate that servant drags Lee's body onstage and rifles his pockets while delivering a monologue in an Irish dialect. The phonetics used to suggest the pronunciation of Mac Chane's lines match those of O'Neale and MacMorris. The character's costume also matches the dress of the Irish natives as they were described in contemporary British histories. When Mac Chane robs Oldcastle's loyal English servant of his clothing, he leaves him with "nothing but a lowsie mantle and pair of broags."[53] Finally, the Irishman refutes the authority of his English overlords. After Mac Chane is apprehended by the constable, the Mayor of Hereford questions him about Lee. Mac Chane responds, "Be Sent Patrick I ha no Mester."[54] Unlike MacMorris, who rejects his national origins in the hope of being counted among King Henry's loyal officers, Mac Chane denies that he has an English master and shows little regard for English laws.

The play concentrates on the legal aspects of imperial expansion and the power of the law to promote religious reform. It begins with the assize at Hereford and ends with a courtroom scene at the same outpost. The scene includes two servants: Harpool, Oldcastle's loyal English attendant, and Mac Chane, Lee's murderous Irish servant. The costumes and accents of these characters initially confuse the court. Mac Chane had slept alongside Harpool in a stable and robbed him of his livery. As the judge explains, Harpool appears "in habite Irish, but in speech not so," while Mac Chane's "speech is altogether Irish, but in habite Seemes to be English."[55] The judge's examination reveals the Irishman had attempted to mask his true identity in order to hide his villainy. When the judge demands his name, he replies, "Me be Mac Chane of Ulster."[56] Unlike MacMorris, Mac Chane is not an Anglicized Old Englishman. He is from Ireland's wildest and most intractable region and a relative of Shane O'Neill, the same notorious Old Irish rebel depicted in *Captain Thomas Stukeley*.

The constable of Hereford identifies Mac Chane as "this savage villain, this rude Irish slave." Lee's father further alienates the servant, calling him a wolf and a snake. Unlike an ass or puppy, the domesticated but ignorant animals that Fluellen uses to describe MacMorris, wolves and snakes were wild and dangerous beasts that, according to British historians, haunted pagan Ireland. When Essex prepared to leave for Ireland and defeat Hugh O'Neill, Thomas Churchyard wrote that the Irish had the minds of wolves and knew no rule of law. He prayed that, with God's help, Essex would punish these devils.[57] Like Alexander Mack Surlo in *Captain Thomas Stukeley*, the judge in *Sir John Oldcastle* uses Mac Chane's punishment as an example of how to weed out the wicked savages infesting Great Britain and Ireland. When Mac Chane admits to murder, the judge orders him to be executed and his body hung in chains. Mac Chane does not ask for redemption. He mutters that would like to be executed in his own clothes and hung according to the Irish fashion.[58] The lesson is simple: there is no place in Great Britain for such a villain, and English authorities must be careful to identify those wicked Irish who feign conformity in order to murder and rob innocent Englishmen.

In this regard, *Sir John Oldcastle* provides a xenophobic Anglo-Protestant history that highlights the ability of law and religion to reinforce racial and cultural boundaries. Rather than reiterating the vision of an inclusive empire found in Shakespeare's *Henry V*, the play presents a nightmare in which the specter of a murderous Irishman moves across the Welsh marches disguised in the habit of an Englishman. Mac Chane's dissembling reveals the anxieties that wracked Englishmen at this time. The character appears to have been inserted into the narrative at a later date. He does not make his entrance until the second half of the play's last act, and he does not appear in Foxe's *Acts and Monuments*, one of its principal sources. Moreover, the play identifies Mac Chane as an O'Neill from Ulster who served an English officer named Lee and describes how they fought the kerns in Ireland before Mac Chane murdered Lee in England. The allusion to prominent figures involved in England's costly war strongly suggests that the play's authors revised their material in light of the latest news coming from Ireland. The playwrights did not complete *Sir John Oldcastle* in the autumn of 1599, giving them time to learn of Essex's failure to defeat Hugh O'Neill in Ireland and his flight with Lee to England.[59]

In 1614, Ben Jonson supplied a more ambitious colonial model through his play *Bartholomew Fair*. The characters in Jonson's play include Whit, Haggis, and Davy Bristle, officers from Ireland, Scotland, and Wales. Their language and actions play upon regional stereotypes, and Whit's malapropisms and aggression identify him as an Irishman.[60] Yet he claims that he is a faithful subject, "Behold man, and see, what a worthy man am ee!"[61] Whit is one of the "Watchmen Three" who help an English captain keep the peace in London. Though Whit's behavior and accent are a source of humor, he does not suffer the punishment that Busy and Wasp, hypocritical English Puritans, suffer under the other watchmen.[62] Whit and his fellow watchmen carry weapons in London and have the power to punish the city's disruptive dissenting Protestants.

The comic band of brothers is similar to the loyal captains in Shakespeare's *Henry V*. Both plays recognize the value of incorporating subjects from the nations bordering England, but Jonson's comedy provides a more radically inclusive scheme. While Shakespeare allows only his Irish captain to fight for the British in France, Jonson uses an Irishman to police English subjects living in London and permits him to court the aristocratic English women visiting its fair. Dame Overdone tells him that she loves "men of war and the sons of the sword."[63] Whit attempts to seduce Win Littlewit and explains that if she leaves her English husband he could make her one of the Lord Mayor's green women and dress her in a green gown.[64] Moreover, the treatment of Busy and Wasp under the authority of the "Watchmen Three" marks the temporary ascendancy of conforming Irish, Scots, and Welsh subjects over Great Britain's dissenting Protestants, suggesting that they are more welcome in Great Britain than the religious radicals among England's dissenters.

While Jonson's play mocks the hypocrisy of London's puritans, it provides an alarming vision of upwardly mobile aliens occupying positions of power at the very heart of Great Britain. This farcical inversion may have alluded to a recent commotion in the Irish government. In May of 1613, the Irish Parliament opened with a symbolic display of unity. The Lord Deputy, peers, noblemen, and conforming clergy processed from Dublin Castle to St. Patrick's Cathedral for special service and then returned to the castle. The most important members of the procession wore scarlet gowns, while the Lord Deputy, Arthur Chichester, appeared in a rich purple robe sent to him by James I. Eight worthy gentlemen bore the Lord Deputy's train as he rode on a fine horse through Dublin's streets. Lord David Barry, Viscount Buttevant, carried the ceremonial sword of state before his Lord Deputy, and Donogh O'Brien, the Earl of Thomond, held the cap of maintenance behind Barry. O'Brien was a loyal Irish officer, but Barry was an Old English Catholic, and he and his fellow recusants would not participate in the Protestant service at St. Patrick's. They waited outside the cathedral and then rejoined the Lord Deputy as he paraded back to the castle quarreling among themselves about who should have greater precedence in the procession.[65]

When the Lord Deputy reached the high house of parliament, he ascended to his seat of state, and the Lord Chancellor made a speech praising King James's newly selected Speaker of the House, Sir John Davis. The recusants objected; they had not voted on Davis, a New English Protestant and former Attorney General. On the second day of the session, the Catholic faction grew more agitated. Davis's supporters could not hear themselves over the tumult, and they left the chamber to determine if they had the numbers to override the Catholics. Once they left the chamber, the recusants locked its doors from within and placed Thomas Everard in the speaker's seat. The loyalists replaced Everard with Davis after two more days of fighting, but Chichester had to adjourn parliament and grant the leaders of the Catholic faction permission to petition King James directly.[66] They traveled to London later that summer, and met with the King at court, but James refused to give Catholics liberty of conscience or remove Protestants like Davis from high office.[67]

By 1615, Chichester instructed his subordinates that King James wanted his officers to cherish and defend Ireland's civil subjects and instruct the disobedient "to embrace knowledge and civility" through justice and good government.[68] The King's generous policies included the sale of English-styled titles in Ireland, which he used to fund other imperial schemes. Many Old English and Old Irish lords purchased such titles in the hope of assimilating into the British social structure that dominated the kingdom by the early seventeenth century.[69] In this respect, while Jonson's play mocks the odd behavior of its outlandish characters, it provides an inclusive colonial model that incorporates loyal subjects from the more remote parts of the British Isles, bringing them closer to the center of the empire. Jonson's model inverts the hierarchy that normally structured British society for comic effect, but unlike the rigid boundaries that separate the "Gentleman of England" from the "Civil Irish Man" and "Wild Irish Man" in Speed's maps, Jonson's play portrays the fluid movement of subjects among the British Isles as both comic and potentially positive. In this regard, his watchmen and Shakespeare's captains are much like Queen Anne's Ethiopian princess. They are considered less civilized than their English counterparts, but they aspire to become British.

Figure 9.3 Irish kern attack an English settler in a plate from John Derricke's *Image of Irelande*, 1581, by permission of the Folger Shakespeare Library.

Conclusion: An Irish Othello

The four plays that I have described in this chapter offer different representations of Ireland's inhabitants and suggest how authorities should position each group within the British Empire. In each play, the language, customs, and manners of the Irish separate them from their British counterparts. The swearing of oaths and the inclusion of other religious references suggest that the performance of these plays demonstrated how certain Irish subjects obstinately clung to backward religious customs that frustrated the attempts of British administrators to reform them. The early printings of these plays also indicate that their performances employed a similar dialect and some used specific costumes and stage business that mimed popular accounts of the Irish to vividly display their brutish manners and belligerent behavior.

The plays reflect evolving and conflicting arguments regarding which of Ireland's subjects could be reformed and which policies would make them more peaceful and industrious. Though these plays portray the Irish as less civilized than their British counterparts, they differentiate between the kingdom's Old Irish and Old English subjects, ranking the latter as more capable of being incorporated into British society. The Old Irish and Old English characters found in these plays share many traits: they use the same dialect and often resort to violence. Yet, while MacMorris and Whit represent loyal Old English officers who deserve to be assimilated into British society, the actions of O'Neale and Mac Chane display the treachery of the Old Irish and warn that they and similarly villainous subjects should be excluded from Ireland's government. The danger was that certain Irish subjects were able to adopt the language, dress, and manners of their British masters and mask their innate villainy.

In 1581, John Derricke warned English administrators to not be deceived by the outward appearance of supposedly conforming Irish natives. He advised British settlers living in Ireland to stay vigilant, "And give not you, your lives to keep, to such dissembling elves."[70] After O'Neill was defeated in 1603 and the 1641 uprising was finally suppressed, British historians used similar language to describe Ireland's rebellious subjects. The histories describe the two rebellions as horrid tragedies enacted on Ireland's bloody stage. In this dramatic accounts, Ireland's apparently obedient Old Irish and Old English lords play the part of treacherous villains.[71] The rebel Catholics mask themselves "under the vizard of submission," or wrap themselves in "the cloak of subjection," and feign their loyalty to the English monarchy.[72] The British histories claim that such dissimulation could not hide the savage nature of Ireland's backward natives. Writing in 1646, Sir John Temple argued that the Irish were still a barbarous people and no law, constitution, or public benefit could draw them from their hatred of the English.[73] Mac Chane's actions in *Sir John Oldcastle* anticipate Temple's concerns and provide a dramatic enactment of Derricke's earlier warning. He takes on the cloak of subjection only to murder his master and later dresses in Harpool's clothes to pass as an Englishman. In *Captain Thomas Stukeley*, O'Neale also claims that he will submit to the Queen's deputy and become a loyal servant after he fails in his attempt to murder the English forces at Dundalk.

The plays with Irish characters also pair different figures in ways that affirm the union between the British monarchy and its loyal subjects. The success of this marriage depended upon the race, culture, and religion of the person in power, and, in 1599, Queen Elizabeth and her conforming clergy claimed that God gave them the authority to rule the Irish.[74] As a result, the Irish characters in these plays do not make a suitable match for a conforming Protestant woman or have legitimate authority over a British subject. Near the end of the sixteenth century, Spenser argued that Old Irish and degenerate Old English lords should not be trusted to raise a ward of court or the children of British nobles. He feared that the children would learn the language and customs of their Irish masters and adopt their lewd behavior.[75] After the Ulster uprising, Sir John Temple claimed that the rebellion was the consequence of the Irish and English intermarrying and making "as it were a mutual transmigration into each others manners."[76] According to Temple, blurring the distinctions that separated the kingdom's British settlers from the island's natives made both groups less civil and obedient. After the Restoration, Sir William Petty proposed a solution. He claimed that he could reform Ireland through a process of transmutation. Petty's plan would send eligible Irish women to England where they would marry Englishmen, learn their language, and conform to the religious practices of their masters. This would allow them to produce loyal subjects who could return to Ireland and make it a stable and industrious kingdom.[77]

These warnings make Dame Overdo's attraction to Whit and his advances to Win Littlewit especially ridiculous. British historians from this period would argue that a match between Jonson's Irish captain and an English woman would be inappropriate, however loyal Whit may be. MacMorris's sway over the Duke of Gloucester, King Henry's youngest brother, is equally inappropriate. The same logic highlights the unnatural nature of the relationship between Mac Chane and Harpool. The two servants lie together in a stable, and when Harpool wakes, he appears to have been translated into an Irishman and is mistakenly apprehended as a criminal while the murderous Mac Chane roams the Welsh marches pretending to be an Englishman. Despite the different models that each play provides, they did not signal a serious revision of Great Britain's social structure. Conservative Protestant leaders did not advocate for policies that mirrored the inclusive models found in Shakespeare's *Henry V* and Jonson's *Bartholomew Fair*. Plays like *Captain Thomas Stukeley*, *Sir John Oldcastle*, and *Othello* better express the colonial schemes that these leaders promoted. These plays display the dangers of miscegenation or giving people of different races, cultures, and religions power over Christian Europeans. By concentrating on the dissembling Old Irish character in *Sir John Oldcastle*, it appears that the Lord Admiral's Men dramatically displayed the dangers of intermarriage and cultural hybridity. In this regard, the play repudiates policies that sought to incorporate loyal Irish subjects into the government that dominated Ireland and Great Britain during the period in which the play was first performed.

The references to sixteenth-century Irish history in *Sir John Oldcastle* suggest that Hugh O'Neill is the play's real villain. He was the brother of the infamous Ulster rebel, Shane O'Neill, Mac Chane's namesake. After Hugh O'Neill's father died and his brother was assassinated, English administrators made O'Neill a ward of the court. Giles Hovendon, an important New English settler, raised him, making sure that he received an English education. The Anglicized O'Neill traveled to London on more than one occasion, hoping to regain his father's English title. He commanded a troop of horse in the service of Queen Elizabeth and fought against Irish rebels for the first Earl of Essex between 1573 and 1575.[78] The Queen once called him "a creature of our own."[79] He even married Mabel Bagenal, the sister of Sir Henry Bagenal, an English administrator who competed with O'Neill for the control of Ulster. Bagenal did not condone the match, and O'Neill married his sister in a secret ceremony.[80]

O'Neill had enemies who feared his influence over Ulster's Gaelic lords. He was an assimilating Old Irish aristocrat who commanded English forces and maintained close ties with other powerful Gaelic families. His different affiliations allowed him to move between Ireland's Anglo and Gaelic communities with ease, gathering his strength in Ulster as he maneuvered through each, but near the end of the sixteenth century he was proclaimed a traitor and joined a general rebellion in the name of religious freedom. O'Neill and his fellow Irish lords were not defeated until 1603, but by that time they had allied themselves with the Spanish who landed an army at Kinsale to support the rebels against the English.

In his upbringing and early military service, O'Neill was much like Shakespeare's Othello, and Andrew Hadfield argues the play has a "ghostly Irish context."[81] This is not to say that Shakespeare wrote the tragedy in response to O'Neill's rebellion. Rather, from the perspective of British authorities in the early seventeenth century, the Moor and O'Neill appear to follow a similar tragic path. The fictional character and historical figure were raised as wards and adopted the language and culture of their leaders. Each attempted to prove his loyalty by fighting the enemies of his master, and both secretly married into the society of their superiors. Yet, according to colonial discourse found in the histories and plays written in early modern England, the outward appearance of these assimilated servants masked an inner darkness. Such narratives helped to portray the Irish as especially backward and savage. According to Derricke, no court training or religious reforms could dispel the bog from the Old Irish as "wildnesse is their fore-possessed state."[82] Such essentialist arguments employed physical and cultural characteristics to link the uncivil natives of Ireland and Africa, much like the pairing of O'Neale and Muly-Mahamet in *Captain Thomas Stukeley*. Similar narratives position such characters as inferior and dangerous aliens when compared to the civilized inhabitants of Great Britain. A small number of writers and administrators contested this colonial discourse, arguing that certain aliens could be reformed. However, many others depicted the Irish as an uncivilized race and argued that their incorporation threatened to undermine an orthodox national identity.

Notes

1. Joseph Roach identifies the anthropological and taxonomic functions of the Augustan theater. He argues that new technologies allowed theater artists to create settings and costumes to better indicate specific locations and their exotic inhabitants during the late seventeenth century. In this way, the proscenium theater operated like a microscope to render visible "an abundance of strange and amusing objects, localities and cultures." My analysis is based on Roach's research. However, I argue that the rehearsal of certain racial stereotypes allowed theater artists in early modern London to address important topical issues without having to represent specific locations through elaborate scenic devises. Joseph Roach, "The Artificial Eye: The Augustan Theater and the Empire of the visible," in Sue-Ellen Case and Janelle G. Reinelt, eds. *The Performance of Power: Theatrical Discourse and Politics* (Iowa City: University of Iowa Press, 1991), 135–136, 142–143.
2. Bernadette Andrea, "Black Skin, the Queen's Masques: Africanist Ambivalence and Feminine Author(ity) in the Masques of Blackness and Beauty," *English Literary Renaissance* 29, no. 2 (1999): 256, 263.
3. Emily C. Bartles, "*Othello* and the Moor," in Garrett A. Sullivan Jr., Patrick Cheney, and Andrew Hadfield, eds. *Early Modern English Drama: A Critical Companion* (New York: Oxford University Press, 2006), 140.
4. *Othello*, act 1, scene 3, lines 1–239.
5. Bernadette Andrea, "The Ghost of Leo Africanus from the English to the Irish Renaissance," in Patricia Clare Ingham and Micelle R. Warren, eds. *Postcolonial Moves: Medieval through Modern* (New York: Palgrave Macmillan, 2003), 203.
6. *Othello*, act 1, scene 1, lines 110–113, 133.
7. Though he attacked Shakespeare's play eighty years after its first performance, Rhymer's xenophobic argument is worth quoting at length: "the character of that state [Venice] is to employ strangers in their wars, but shall a poet fancy that they will set a Negro to be their general, or trust a Moor to defend them against the Turk? With us, a Blackamoor might

rise to be a trumpeter, but Shakespeare would not have him less than a lieutenant-general. With us, a Moor may marry some drab or small-coal wench; Shakespeare would provide him the daughter and heir of some great lord or Privy Councillor." Thomas Rymer, *A Short View of Tragedy* (London, 1683); repr. in *The Critical Works of Thomas Rymer*, ed. Curt Zimansky (New Haven: Yale University Press, 1956), 132–133.

8. Andrea, "Black Skin," 268.
9. Kim Hall, "Sexual Politics and Cultural Identity in the Masque of Blackness," in Sue-Ellen Case and Janelle G. Reinelt, eds. *The Performance of Power: Theatrical Discourse and Politics*, (Iowa City: University of Iowa Press, 1991), 54.
10. Ben Jonson, *The Masque of Blackness* (London: 1605); repr. in *The Works of Ben Jonson* (Boston: Phillips, Sampson, 1853).
11. *The History and Topography of Ireland by Gerald of Wales*, trans. John O'Meara (London: Penguin Books, 1982), 31.
12. Ibid., 101.
13. Ibid., 69–76, 110.
14. According to Raymond Williams, authors first used the term "community" in the sixteenth century to describe a group that shared the same beliefs, values, and interests, as well as other special characteristics that allowed each community to function as a separate group within a larger political structure, *Keywords: A Vocabulary of Culture and Society* (New York: Oxford University Press, 1976), 75–76. Sixteenth- and seventeenth-century British historians argued that the differences that separated Ireland's three communities were distinct and determined the essential character of each subject. For the terms Old Irish, Old English, and New English, and how a person identified such differences among each group, I have relied on the definitions found in Richard Cox's *Hibernia Anglicana* (London: 1689). Though William Camden's Latin edition of *Britannia* (London: 1590) was published almost a century before Cox's account, Cox uses the same terms as Camden to identify Ireland's different subjects earlier in the sixteenth century.
15. Edmund Campion, *The Historie of Ireland, collected by three learned authors viz. M Hanmer, E. Campion, and E. Spenser*, ed. Sir James Ware (Dublin: Printed by the Society of Stationers, 1633), 14; William Camden, *Camden's Britannia, Newly Translated into English with Large Additions and Improvements* (London: Printed by F. Collins, for A. Swalle at the Unicorn at the West end of St. Paul's Church-yard; and A&F Churchill at the Black Swan in Pater Noster Row, 1695), 148; Fynes Moryson, *The Rebellion of Hugh Earle of Tyrone, and the appeasing thereof; written in the form of a journal. An itinerary written by Fynes Moryson Gent* (London: Printed by John Beale, 1617), 2.
16. Moryson, *Rebellion of Hugh Earle of Tyrone,* 300.
17. Robert Payne, *A Brife description of Ireland: Made in this yeere, 1589* (London: Printed by Thomas Dawson, 1590).
18. Ibid., 4, 13.

19. Thomas Lee, *A Brief Declaration of the Government of Ireland* in *Desiderata curiosa Hibernica*, ed. David Hay (Dublin: Printed by David Hay, 1772), 89, 94.
20. Ibid., 131.
21. Edmund Spenser, *A View of the State of Ireland* (Malden, MA: Blackwell, 1997), 15–18.
22. Ibid., 27, 36, 143.
23. Camden, *Camden's Britannia,* 993.
24. Campion complained that the native Irish were at fault: "The very English of birth, conversant with the brutish sort of that people, become degenerate in short space, and are quite altered in the worst ranke of Irish Rogues" (14). Edmund Spenser agreed with this assessment, but he also complained that there is an "evil that hangeth upon that countrey.." This evil included the backward customs of the island's uncivil inhabitants and its poor climate. These factors contributed to Englishmen forgetting their natures and forgoing their nation (Spenser, *View of the State of Ireland,* 12, 54.).
25. Camden, *Camden's Britannia,* 117.
26. Ibid., 141–143, 148.
27. Moryson, *Rebellion of Hugh Earle of Tyrone,* 300.
28. Sir John Temple, *The Irish Rebellion: With the Barbarous Cruelties and Bloody Massacres Which Ensued Thereupon* (London: Printed by R. White for Samuel Gellibrand, 1646), 14.
29. Richmond Barbour explains the similar mimetic functions in each medium: "Given the master trope of the world as a stage, a book of maps can be a theater, and a playhouse can be a three-dimensional map of the world," Richmond Barbour, "Britain and the Great Beyond: The Masque of Blackness at Whitehall," in John Gillies and Virginia Mason Vaughan, eds. *Playing the Globe: Genre and Geography in English Renaissance Drama* (Madison, NJ: Fairliegh Dickenson University Press, 1998), 129.
30. Camden, *Camden's Britannia,* 110.
31. John Speed, *Theatre of the Empire of Great Britain* (London: 1611), 137–138.
32. The images of the "K. of Florida" and a "Woman from Mexico" appear above one another in a cartouche inserted into the two-leaf map of America. Speed, *Theatre of the Empire of Great Britain,* 8–9.
33. Atlases that popularized images of Africa in the seventeenth century employed the same format that structures the cartouche in Speed's map of Ireland. In each case, the arrangement of the characters creates a hierarchy that positions whiteness above darkness. The characters at the top of the two cartouches inserted into Bleau's map of Africa are whiter, wear more clothing, and appear more civilized. The natives at the bottom of each cartouche are dark, mostly naked, and wear exotic feathers and skins.
34. Anonymous, *The Famous History of the Life and death of Captaine Thomas Stukeley. With his Marriage to Alderman Curteis Daughter, and valiant ending of his life at the Battaile of Alcazar* (London: Printed for Thomas Pauyer, 1605).

35. Camden had published a Latin edition of *Britannia* as early as 1586. Philémon Holland published an English translation of Camden's work in 1610. Derricke published the *Image of Ireland* in 1581; John Derricke, *The Image of Irelande* (London: 1581; repr. by John Small, Edinburgh: Adam and Charles Black, 1883).
36. The Lord Chamberlain's Men most likely staged the play at the Curtain, an older theatre located in Shoreditch just north of the city. James Shapiro, *A Year in Life of William Shakespeare, 1599* (New York: HarperCollins, 2005), 78–80; 88–89.
37. *Henry V*, act 3, scene 2.
38. Giraldus describes Ireland as disjointed and distinct from "Greater Britain" Giraldus, "The History of the Conquest of Ireland"; Camden argues that no other group of islands were "greater, larger, or better, than the British Isles," and the "Britans" were separate and superior race. According to Camden, after the Romans conquered the "Britans," they named Great Britain, Britannia Prima, and Ireland, Britannia Secunda (Camden, "Annales"). Though the Roman names indicate the size of the islands, Elizabethan and Jacobean historians used the terms to accentuate Great Britain's dominance over the other islands in the archipelago.
39. Margot Heinemann, "Political drama," in A.R. Braunmuller and Michael Hattaway, eds. *The Cambridge Companion to English Renaissance Drama* (Cambridge: Cambridge University Press, 2003), 177.
40. Johnson wrote his jingoistic hagiography between 1596 and 1597, two years before the Lord Chamberlain's Men performed Shakespeare's *Henry V*. The narrative was adapted into different plays, including John Kirk's *Seven Champions of Christendom* (London: Printed by J. Okesn 1638). Kirke's play was staged at the Cockpit and Red Bull before the interregnum (Gerard Langbaine, *An Account of the English Dramatic Poets* [Oxford: Printed by L.L. for George West and Henry Clements, 1691], 84). Mummers also incorporated Johnson's characters into their plays (J.R.R. Adams, *The Printed Word and the Common Man: Popular Culture in Ulster, 1700–1900* [Belfast: Institute of Irish Studies, Queen's University of Belfast, 1987], 4). In the seventeenth century, English and Scots settlers brought these plays to Ireland. In some of the Irish mummers's plays St. Patrick and St. George fight one another. The result of the mock combat depended on the location of the performance and the composition of its audience. When the native Irish performed their folk dramas in the manor home of a Protestant settler, St. George defeated St. Patrick. In Catholic homes, St. Patrick killed St. George (Alan Gailey, *Irish Folk Drama* [Cork: Mercier Press, 1969], 8, 44–49, 84).
41. Shapiro, *Year in Life of William Shakespeare*, 257.
42. Thomas Churchyard, Esquire, *The Fortunate Farewell to the most Forward and Noble Earl of Essex* (London: Printed by Edm. Bollisant at the west door of Powles, 1599).
43. Giraldus, "The History of the Conquest of Ireland", 104–105.
44. Camden, "Annales", 110.

45. Shapiro, *Year in Life of William Shakespeare,* 101.
46. Unfortunately, though the records indicate that Philip Henslowe paid his playwrights to write a second part of the play, there is no evidence of it ever being printed or produced as a manuscript. *Henslowe's Diary,* ed. R.A. Foakes (Cambridge: Cambridge University Press, 2002), 125, 132; James J. Marnio, "*William Shakespeare's* Sir John Oldcastle," *Renaissance Drama* 30 (1999–2001): 125–126.
47. R.A. Foakes, "Playhouses and Players," in A.R. Braunmuller and Michael Hattaway, eds. *The Cambridge Companion to English Renaissance Drama* (Cambridge: Cambridge University Press, 2003), 6.
48. R.S. Forsythe, "Certain Sources of Sir John Oldcastle," *Modern Language Notes* 26, no. 4 (April 1911): 104.
49. Peter Corbin and Douglas Sedge, *The Oldcastle Controversy: Sir John Oldcastle, Part I and the Famous Victories of Henry V* (Manchester: Manchester University Press, 1991), 11.
50. Ibid., 1–2.
51. Ibid., 2; Shapiro, *Year in Life of William Shakespeare,* 10.
52. Mary G.M. Adkins, "Sixteenth-Century Religious and Political Implications in Sir John Oldcastle," *Studies in English* (1942): 96.
53. *The Life of Sir John Oldcastle,* repr. The Malone Society Reprints (London: Charles Whittingham Chiswick Press, 1908), act 5, scene 7.
54. Ibid., act 5, scene 6, line 2201.
55. Ibid., act 5, scene 10, lines 2510–2513.
56. Ibid., act 5, scene 10, line 2516.
57. Churchyard, *Fortunate Farewell, passim.*
58. *Sir John Oldcastle,* act 5, scene 11, lines 2643–2670.
59. Shapiro, *Year in Life of William Shakespeare,* 10.
60. When Whit speaks the letter "d" replaces "th," "sh" replaces "s," and "ph" replaces "wh." This carefully constructed orthography matches the dialects for O'Neale, Mac Chane, and MacMorris. For example, Whit states "for tee an' tou art in an oder 'orld" (act 3, scene 1, line 3). Haggis has the accent of a Scot, and Wasp calls Davy Bristle a Welsh cuckold who stinks of leeks (act 4, scene 6, lines 41–46); Ben Jonson, *Bartholomew Fair,* in Eugene W. Waith, ed. *The Yale Ben Jonson* (New Haven: Yale University Press, 1963).
61. When Whit takes the stage in the third act, he boldly introduces himself using lines from a popular mummer's play: *Behold, man, and see, what a worthy man am ee! / With the fury of my sword and the shaking of my beard, / I will make ten thousand men afeared* (act 3, scene 2, lines 116–119). The italics appear in the original printing of Jonson's play and draw attention to their lyrical quality. The captain of the mummers used similar lines to begin the folk dramas in Ireland (Waith, *Yale Ben Jonson,* 95; Gailey, *Irish Folk Drama,* 1–3, 55).
62. Waith, *Yale Ben Jonson,* 95.
63. *Bartholomew Fair,* act 4, scene 4, lines 210–211.
64. Ibid., act 4, scene 5, lines 87–88.

65. William Farmer, *A Chronicle of Lord Chichester's Government of Ireland; Containing Certain Chroniculary Discourses for the Years of our Lord 1612, 13, 14, &15*, in David Hay, *Desiderata curiosa Hibernica* (Dublin: Printed for David Hay, 1772), 1:166–167.
66. Ibid., 168–169; 207–207.
67. Ibid., 210–211.
68. Arthur Chichester, "Instructions for the Lord President and Council of Munster, 20 May, 1615" (from Arthur Chichester and his council), in David Hay, *Desiderata curiosa Hibernica* (Dublin: Printed for David Hay, 1772), 2:1–2.
69. In 1611, James I created the hereditary Order of Baronets to help settle Ireland and raise funds for the court. In 1622, Oliver Tuite, a wealthy Old English landholder in Westmeath, purchased the title and became Sir Oliver Tuite of Sonnagh (Edward Kimber, *The Peerage of Ireland* [London: 1768], 227). The Tuites of Sonnagh cultivated their English culture as a means to secure their power, and one of Oliver's relatives, Jane Tuite, married Francis Edgeworth, a New English lawyer. Her portrait appears in *The Black Book of Edgeworthstown*, ed. Harriet Butler and Harold Butler (London: Faber and Gwyer, 1927). Her fashionable dress and Flemish lace provide a stark contrast to Thomas Lee's costume in Gheerhaerts's painting. The portraits blur the distinctions that identified English and Irish subjects during the period and reveal the desire of their sitters to affiliate themselves with different cultures.
70. John Derricke, *The Image of Irelande* (London: 1581; repr. ed. John Small, Edinburgh: Adam and Charles Black, 1883), 40.
71. In his account of O'Neill's rebellion, Thomas Stafford claims that Munster "was the stage whereon the last and greatest scene of that Tragedie was acted" (Sir Thomas Stafford, *Pacata Hibernia, Ireland appeased and reduced: or, an historie of the late warres* [London: Printed by A. Mathewes, 1633], A2). James Cranford used the same metaphor to introduce his graphic account of the Ulster uprising as "this horrid Tragedy, now to be acted on the stage of *Ireland*," James Cranford, *The Teares of Ireland* (London: Printed by A.N. for John Rothwell, 1642), 7.
72. Camden, "Annales", 115; Stafford, *Pacata Hibernia*, 1.
73. Sir John Temple, *The History of the Irish Rebellion* (London: Printed at St. Paul's Churchyard, 1646), 8.
74. In 1599, Queen Elizabeth justified Essex's invasion of Ireland by arguing that kingdom's less civilized rebels had forgotten the obedience that they owed to their natural leaders, and they unnaturally resisted divine ordinance by fighting God's anointed monarch. *The Queenes Majesties Proclamation declaring her princely resolution in sending over of her Army into the Realme of Ireland* (London: Printed by the Deputies of Christopher Barker, Printer to the Queenes most excellent Majestie, 1599); *A Prayer for the good success of her Majesties forces in Ireland* (London: Printed by the Deputies of Christopher Barker, Printer to the Queenes most excellent Majestie, 1599).

75. Spenser, *View of the State of Ireland,* 36.
76. Temple, *History of the Irish Rebellion,* 14.
77. Sir William Petty, *A Political Survey of Ireland: The Political Anatomy of Ireland, 1672* (repr., Dublin: 1689), 29–31.
78. Hiram Morgan, *Tyrone's Rebellion: The Outbreak of the Nine Year's War in Tudor Ireland* (Suffolk: Boydell Press, 1999), 93.
79. University Library Cambridge, Kk 1 15 no. 31, fos. 87–88, in ibid., 85.
80. Ibid., 79.
81. Andrew Hadfield, "'Hitherto she could ne're fancy him: Shakespeare's 'British' Plays and the Exclusion of Ireland," in Mark Thornton Burnett and Ramona Wray, eds. *Shakespeare and Ireland: History, Politics, Culture* (London: Palgrave Macmillan, 1997), 56–60.
82. Derricke, *Image of Irelande,* 42.

CHAPTER 10

Icons of Atrocity: John Derricke's *Image of Irelande* (1581)

Vincent Carey

The bitter lessons of his years of dedicated service to the queen in Ireland dominated Sir Henry Sidney's March 1, 1581/1582 memorandum to Sir Francis Walsingham, Elizabeth's then secretary of state and a key player in her administration.[1] Sidney's "Memoir" of his years of service from 1556 to 1578 has been termed by its most recent editor as "A Viceroy's Vindication?" an apt title for a work that attempted to explain the author's perception of a double injustice involving near financial ruin and a badly damaged political reputation. Sidney's "Vindication" sought to reassure his son's prospective father-in-law that his current straightened circumstances were not the result of a lack of political virtue or failure, but, on the contrary, the result of vicious backbiting on the part of rivals at court, the intrigues of an upstart English-Irish or Old English gentry, and finally the ingratitude of an all too fickle monarch. The last cause was made all the more painful because of the clear evidence, at least from Sidney's perspective, of his multiple initiatives and successes on behalf of his queen.

Central to this personal narrative is an account of rebellions crushed, recalcitrant outlaws brought to boot, and other potential outbreaks nipped in the bud through the nuanced combination of negotiation and force. Like Julius Caesar, whose *Commentarii de Bello Gallico* (Commentaries on the Gallic War) is a major influence, according to Ciaran Brady, on the text, Sidney describes how he came, he saw, and he conquered Ireland. Yet unlike his literary exemplar, the lord deputy returned home in a cloud of criticism to suffer the indignities of a house bankrupted by the expenses of service. A house so straightened that he had to write to his son's putative father-in-law to explain his lack of resources for his son's marriage portion, all the more galling when the queen advanced his rivals, especially, Thomas

"Black Tom" Butler, the tenth earl of Ormond. Sidney's "Memoir" is then part political justification, part "anti-Ormond" rant, part constitutionalist justification for royal "absolutism," and a sustained argument for the application of necessary force. Sidney, like his Roman Proconsul and literary mentor, wanted to demonstrate that he could put down what he saw as cruel and violent rebellion and like Caesar he understood the effectiveness of punitive force utilized in the interest of what Sidney termed *Secundum Jus Talionis*, following violence with punishment in kind, or "retributive justice."[2]

This chapter argues that the concept of retributive justice was central to Sidney's worldview, which depended upon several preexisting ideas, including the preeminence of English monarchical authority, the accuracy of English constructs of justice, and the conviction that only English force applied with diligence and perseverance could bring rogue Ireland into the "civil" fold. This model of the world permitted the English to continue their imperial efforts in Ireland without need for further justification for their forceful approaches to bringing the Gaelic Irish and the Old English into line. Justice according to the English meant the peaceful and appreciative acceptance of their rule, complete adherence to their laws and religion, and the use of any means necessary to accomplish their goals. Ireland was the first, and arguably the most recalcitrant of England's colonial projects. Beginning in the twelfth century an emerging Anglo-Norman kingdom attempted to settle the island next door and bring its inhabitants to bear. The struggle proved far more difficult and prolonged than anticipated. The worldview that emerged from their long-term struggles with the Irish supported their imperial mission in other parts of the world, especially in the Americas. The broader European experiences outside of Europe also supported this attempt to subjugate the Irish, and Sidney's "Memoir" reflects a particular moment in the narrative the English crafted about the Irish, and thus about their definitions of civility and savagery and the relationship between justice and force in an imperial context.

What is most surprising about the exposition of punitive force, in particular, is not the fact that it came at the culmination of the memoir, but that it was illustrated by the story of the hunting down and killing of the relatively minor, though effective, Midland Gaelic lord, Rory Óg O'More. From 1571 until his death in 1578, O'More was the "chief of his sept" of the O'Mores and the extremely effective leader of the Gaelic resistance to the plantation of the Midland districts of Laois and Offaly. By the mid-1570s O'More's guerilla activities had become a serious problem for the settlers and for the government. Unwilling to placate O'More's claims to lordship of the region and determined to preserve the fledgling settlement, Sidney decided on the brutal suppression of Rory Óg.[3] Ciaran Brady in his excellent edition of Sidney's "Memoir" refers to this Rory Óg episode as the "epitome of the entire text."[4] The highlighting of the hunting down of the lord of the O'Mores is still an odd choice for Sidney given the serious issues at stake in the "Memoir" overall. Unless it is the perfect example, and one of his few successful exploits, that promote Sidney's version of power and force in the extension of government, his view that successful military campaigns, coercion—even to the eradication of the enemy and his support network—were the most effective means to bring about the "reform" of the Irish.[5]

Yet surprisingly, the "Memoir" was not the only Sidney apologia to prominently feature Rory Óg O'More. He takes center stage in John Derricke's 1581 "The Image of Irelande with a Discoverie of Woodkarne," one of a handful of works on Ireland to be published in England at the time. Most widely know for its woodcut series, these images today adorn nearly every illustrated book on Ireland in the early modern period.

Yet despite the ubiquity of these reproductions, Derricke's *Image of Irelande* itself has not been, at least until recently, the subject of extensive scholarly attention. This academic interest began at the end of the 1990s with two separate articles, one by the author, and another by Maryclaire Moroney (1999),[6] both of which addressed the extended verse sequence " The Image of Irelande" written at the end of 1578, and placed emphasis on its apocalyptic and anti-Catholic intent. In addition, Moroney wanted to emphasize it as a literary precursor to Spenser's "A View of the Present Sate of Ireland," completed by Edmund Spenser sometime around 1596,[7] while Carey attempted to connect Derricke's doggerel to a brutal campaign waged by Sidney in the Gaelic Midlands in the 1570s. Ciaran Brady's aforementioned edition of Sidney's "Memoir" also brought attention to Derricke's text, but Brady questioned Carey's evidence that purported to associate Sidney with an "espousal of radically violent and even genocidal policies toward the native Irish."[8] Willy Maley's recent essay "Remembering and dismembering in Sir Henry Sidney's Irish *Memoir*" (1583) reiterates the association between Sidney and radical violence, and again draws on Derricke's text as a key piece of evidence.[9] While it is too early to say where this line of inquiry will lead, it needs to be noted that all the above commentators have focused on the verse or text while largely ignoring the woodcut sequence that is part of the work and yet has a separate title page and publication details.

This neglect has been somewhat addressed with the 2003 publication of James Knapp's *Illustrating the Past in Early Modern England: The Representation of History in Printed Books*, a work that places Derricke's woodcuts at the forefront of English illustration history, sharing the limelight with John Fox's *Book of Martyrs* and the printer John Day, whose shop was responsible for both.[10] Knapp suggests that "Derricke's Image of Ireland . . . is perhaps the most dramatic example of the interconnection between history writing and artistic representation in an English book from the sixteenth century."[11] Knapp does not stop here; he wants

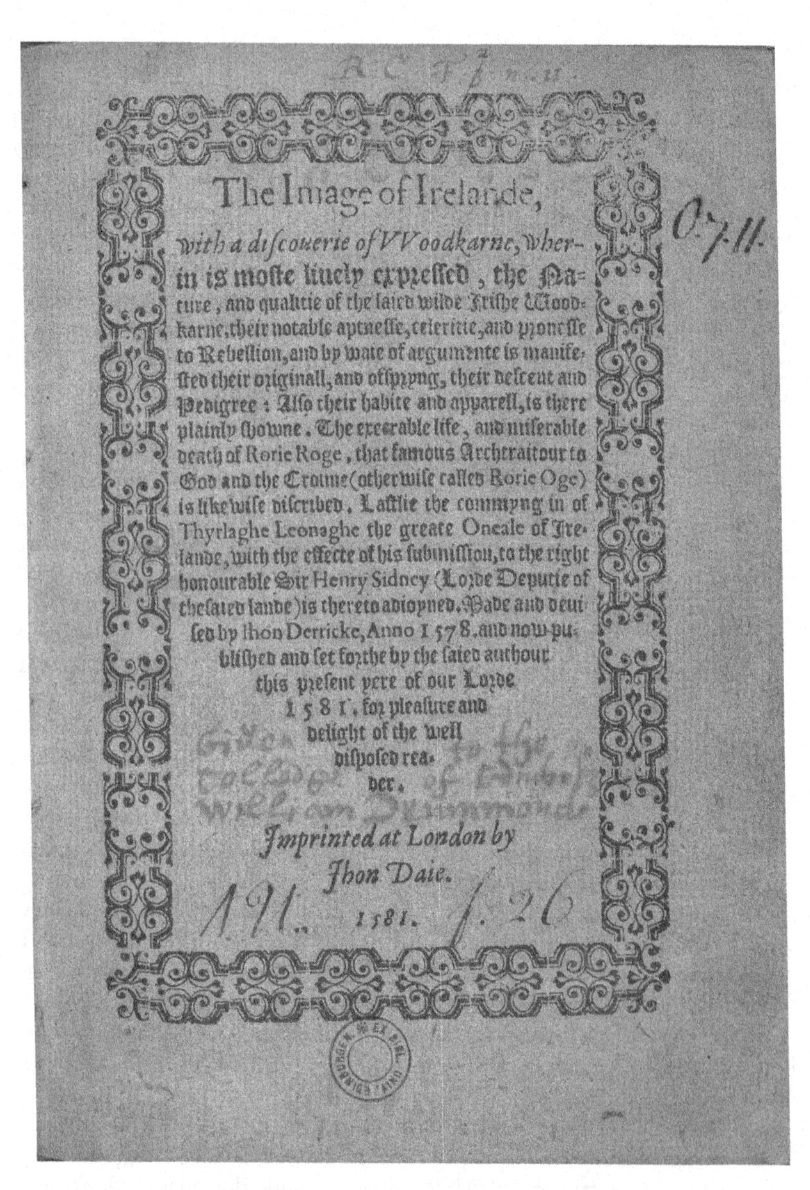
The Image of Irelande,

with a discouerie of VVoodkarne, wherin is moste liuely expressed, the Nature, and qualitie of the saied wilde Irishe Woodkarne, their notable aptnesse, celeritie, and pronesse to Rebellion, and by waie of argumente is manifested their originall, and offspryng, their descent and Pedigree: Also their habite and apparell, is there plainly showne. The execrable life, and miserable death of Rorie Roge, that famous Archtraitour to God and the Crowne (otherwise called Rorie Oge) is likewise discribed. Lastlie the commyng in of Thyrlaghe Leonaghe the greate Oneale of Irelande, with the effecte of his submission, to the right honourable Sir Henry Sidney (Lorde Deputie of the saied lande) is thereto adioyned. Made and deuised by Ihon Derricke, Anno 1578. and now published and set forthe by the saied authour this present yere of our Lorde 1581. for pleasure and delight of the well disposed reader.

Imprinted at London by Jhon Daie.

1581.

Figure 10.1 Title page of John Derricke's *Image of Irelande* (1581). Special Collections Department, Edinburgh University Library.

his readers to understand that "*The Image* can be seen to participate in a shift from a program of reform to one of repression, in a movement facilitated by a concomitant shift in representational strategy from a visual and explanatory mode to one grounded in the verbal concept."[12] Citing Patrick Collinson he asserts that Derricke's text "straddles" a "watershed moment in England's cultural transformation from a culture of 'orality and image to one of print culture: from one imaginative 'set' to 'another.' "[13] These are dramatic claims that warrant serious and extended consideration, not the sort of treatment possible in a short chapter like this.

The remainder of this chapter explores the political worldview suggested by these wood cuts, a cosmogony that asserted that retributive justice and campaigns of violence were essential to bring the Irish into the civilized fold. Furthermore it argues that Derricke's woodcuts provide us with evidence of Sidney's worldview cannot be doubted. It is highly unlikely that as a demonstrated supporter he would promote positions that he thought were out of touch with Sidney's or that he thought would damage his chances of reappointment as lord deputy of Ireland. It is also essential to consider the implications of this message for people on the ground in Gaelic Ireland. For these images tell a tale of the centrality of retributive justice in the English colonial approach, rather than obscure the violence inherent in the colonial process they celebrate a violent fury unleashed, and at the same time advertise its success as a model to be emulated. In fairness, James Knapp does not shy away from the harsh implications of this literary and artistic enterprise. In fact, in noting the distinction between the first six crude and unsigned woodcuts depicting the Gaelic Irish as barbarous and contrasting this with the next six signed woodcuts portraying the "civil" power and discipline of English military might as led in the field by Sir Henry Sidney, Knapp sees a complex representational strategy:

> For in the relation of text to image, we have an example of a thoughtful manipulation of form in the service of a specific political message. While the polemical anti-Irish position is relatively uncomplicated, the justification for the harsh treatment of the Gaelic-Irish relies on an aesthetic understanding of civility, which emerges from the interpretation of the verse and the illustrations.[14]

Knapp certainly does not shy away from the connection between image and violence, although he does suggest that the effort at cultural transformation implied in the overall work, though "coercive," was "compatible with Protestant efforts to reform religious practice in the first two decades of Elizabeth's reign."[15] Here Knapp shares with Brady the view that the ultimate goal of the Sidney government was "reform," that most problematic of terms, and further, the opinion that Derricke's images/woodcuts share with the extended poetic sequences in Parts I and II of the work, a belief in the possibility of political and cultural transformation in Ireland once the problem of the violent elements, categorized as wood kerne in the text, have been dealt with.[16] Knapp argues that Derricke's "strategy still points to conversion as the solution" and "To this end as the poem moves between the poles of Irishness and Englishness, between the wild and the civil, it is punctuated with some diplomatic successes, which mitigate the considerably brutal acts of repression."[17]

I find this assertion doubly problematic as it seems to suggest a unity within the text that is not consistent, and it also seems to distance Derricke's work, and by extension Sidney from the frenzy of violence unleashed in 1578. If we consider, following Richard McCabe, that the use of the term "kerne" is a metonymic for Gaelic culture and society, the woodcuts as a whole clearly represent the worldview, that without radical violent acts against the recalcitrant elements of Gaelic society, the English colonial enterprise will be overthrown and the "true faith" lost to Catholicism.[18] For the Sidney worldview as represented in Derricke's "images" English justice was God's justice and thus its inherent violence was justified in this providential sense.

The link with the Sidneys, father and son, is made clear right from the start of Derricke's text with a dedication to "maister Philip Sidney" referred to as his "benefactor."[19] A client relationship is implied with the note that the work was composed in "discharge of my deutie towards my lorde your father"[20] and accomplished during service with Sidney senior in Ireland. Derricke finished writing the first part of the work, according to the dedication, before June 16, 1578.[21] The text was then augmented to include a report on the death of Rory Óg O'More, which occurred on June 30 and was reported by Sidney in a letter to the council on July 1, 1578.[22] Derricke then wrote the section on Rory Óg after July 1 in a different context and in a sense of celebration of the deputy's success.[23]

The goal was to augment an already completed project and presumably one composed over a longer period of time. The author informs us that he wants for the "volumes augmentation... to put next in sequence, the pictors and protractors of the most notablest Rebelles in Irelande."[24] Yet when it comes time to pick these, the only one he comes up with for his "same substantial cornerstone" is Rory Óg. Derricke's next rhetorical move is most interesting for he decides to ventriloquise to focus on Rory as fugitive, and in the first person, describe Sidney's prosecution, the circumstances that brought him so low. He puts into Rory's mouth the prophecy of his own doom, which Derricke describes as "his fatall destinie."[25] And suggesting that the woodcuts were already completed and before him as he wrote his update, Derricke asks the reader to "behold in plaine ***protactour***, a grose and corpulent man, lapped in a mantel overwhelmed with miserie, beyng in a wood (an ill favored churle) standyinge on a Hillocke."

Figure 10.2 Rory Óg "tombled in the myre," Plate 11 from John Derricke's *Image of Irelande* (1581). Special Collections Department, Edinburgh University Library.

I have written about the next section of the text extensively elsewhere, suffice it to say that Rory is made to repent of his rebellion and bemoan his vile life in sin and "satan[s] . . . thrall" (200). Though his fate may seem inevitable, the real Rory Óg was in fact hunted down and killed in the summer of 1578—the verse is written to suggest that Sidney gave him "pardon" and repeated opportunities to repent—"O hapless wight refusing prince's grace" (201). Hunted by the state's "warlike crew," "Like greedy hawks pursuing fast their prey," Rory can only recognize the inevitability of Sidney's retributive justice; "and who but he might Rory overthrowe, Though mars himself had sworn my mortal foe" (202). Divine retribution justified the brutal campaign of extermination that Sidney had now unleashed, retribution that included the burning of his wife and the killing of many of his children: "With many more my dear and special friends, whose breathless corpses were give to flames of fire" (202). And Rory is made to speak of the elimination of thousands of O'Mores and their allies the O'Connors: "When thousand *Ms* and *Cs* . . . debarred shall for thereof come the ill" (204). Derricke then employs prophecy as Rory is made to foretell the extermination of his people the O'Mores, and not just the kerne it needs to be noted.

> That time draws nigh and hour is at hand,
> In which the sept of my rebelling race,
> Shal be extirped and abolished clean the land
> For God himself do sit in judgment place. (205)

The fictive Rory prognosticates of the "fearful scythe" of divinely ordained retribution: "For God and time hath sworn by sacred oath, that reed and husk shall suffer penance both" (205). The herald of death that Rory speaks of is, of course, Henry Sidney who "by sword and cruel conquest" (206) will bring peace to the land. As noted above, the real Rory was finally killed in June 1578 after a campaign against him and his followers directed by Sidney. Though the lord deputy did not have a hand in his actual demise, Derricke's work makes clear that the campaign between the Gaelic insurgent and the forces of the state was a war to the death, employing methods that if necessary would have "extirped" the entire O'More clan and "abolished clean the land."

The effectiveness of these methods is then literally illustrated in the final portion of the work, the separate woodcut section. These images need to be examined and read in conjunction not alone with the earlier and more "reformist" Parts I and II but with the updates added in the early months of 1578, culminating in Rory's death in 1578. As such they represent a moment of heightened desperation in the Sidney camp and visual resume of the effectiveness of brute force. It needs to be noted that though crude violence is represented here it comes nowhere near the actual violence visited on the midlands from February to June of that year. Sidney's campaign against the O'Mores and O'Connors included the murder of women and children, burning, entrapment, massacre, and the search-and-destroy tactics familiar to us today. These icons of atrocity make little bones about the necessity of these acts of terror and suggest that the target maybe Rory Óg and the kerne, the latter is really "employed as a metonym for Gaelic society generally."[26]

A cursory examination of the first woodcut in the series (Plate 1, figure 10.3), for instance, demonstrates that from the outset Derricke's worldview does not discriminate among the various and complex levels of Gaelic society, tarring all with the same brush. The verse below the image clearly notes:

> Mark me the karne that gripes the axe fast with his murd'ring [h]and, Then shall you say a righter knave came never in the land; As for the rest so trimly dressed, I speak of them no evil, in each respect, they are detect as honest as the devil.

While the verse and the image purport to describe the murderous nature of the wood kerne, the Gaelic elite here "so trimely dressed" are also included and typed as dishonest as the devil. In other words, the Irish, whether leaders or subjects, military personnel or children, are all inherently dishonest and uncivilized, and thus whatever happens to them at the hands of the godly English will be divinely intended and perfectly all right. The difference between the civilized Sir Henry Sidney (as illustrated in figure 10.11 below)[27] and the Irish lord is illustrated by their dress. Sidney, clad in a brocaded coat, tall hat, boots with spurs, and sporting a neatly trimmed beard and moustache, looks like he would easily fit in back at Whitehall, while the Gaelic lord, wearing his mantle and a short coat, is accompanied by his kerne (to the left) who wears only a short shirt that barely covers his genitals and exposes his chest. While the lord wears tights and spurs, the kerne on the left wears no pants or "trews"; the kerne to the right holding the horse wears neither "trews" nor shoes, indicating the distance between the Irish kerne and any hope they might have for civility. Despite the "civilly" dressed Gaelic lord and his elegantly saddled and much favored small breed of horse, the Irish "Hobby," the clear association is with the savage wood kerne.[28]

The second woodcut (*Image*, Plate 2, figure 10.3) in the series continues to build the worldview that all the Gaelic Irish are endemically violent: "Most hurtfull to the English pale and noisome to the states. Which spare no more their country birth, then those of th' English race . . . They spoile, and burn and, beare away . . ." While purportedly representing the kerne raiding a farm in the Pale, the area of Old English settlement and society, the verse and the image (below) visualize Gaelic violence and include a

Figure 10.3 A Gaelic lord, "as honest as the deuill," Plate 1 from John Derricke's *Image of Irelande* (1581). Special Collections Department, Edinburgh University Library.

Gaelic elite figure (on the horse in the center foreground). Again despite the effort to represent the violence as that of the rootless and landless kerne, an image that could resonate with the civil English readers who lived in fear of armies of vagrants and demobbed soldiers wandering the English countryside, the presence of a Gaelic figure on horseback and the inclusion of a piper with rather impressive war pipes or bag pipes suggests a more targeted punitive raid rather than simple thievery. Derricke chooses to represent this cattle raid at the outset of his Sidney resume to show that settled people, even Gaelic farmers will be safer when the fundamentally flawed Gaelic military class and their elite leaders are brought to justice.

That retributive justice is necessary to curb the violence of the preceding image of Gaelic violence is reinforced throughout the following verse and image sections of the work. That the target of this violence is Gaelic society, that is, all the Irish and not just the kerne, is visually represented in the third woodcut in the series (*Image*, Plate 3, below). In the image of the Gaelic lord feasting with his wife and retinue, all the elements of a fundamentally flawed society are represented, from the tonsured Catholic friar who "Now when into their holdes are entred in, . . . And fryer smelfeast sneaking in, doth preace amongst the best," to the harper and the bard and the bare-arsed kerne by the fire. While there are kerne in this image, they are, in fact, marginal. What is central is a Gaelic lord, his wife, and the Gaelic clerical and learned caste. This Mac Suibhne Fanad, or Mac Sweney, the subject of the woodcut, stands for the social customs of the entire society that are typed hog like.

There is also the possibility that the metropolitan English reader may have found this scene with its disemboweled and disembodied animals and cooking utensils reminiscent of images of "New World" cannibals.[29] While it is clear from the accompanying verse that the feast is derived from the stolen cattle, and that the lack of cooking utensils forced them to "plucketh off the Oxes cote, which he euen now did weare, . . . to boyle the flesh his hide prepare," the images nonetheless evoke anthrophomorphic responses. Derricke's denigration of the Gaelic elite, despite the clear sartorial distinction between it and its kerne following, is reinforced by the Latin phases uttered by the two farting kerne (they are *braigetori* of professional farters)[30] on the far right: "Aspice spectator sic me docuere parentes" (see beholder, this is how my parents taught me) and "Me quoque maiores omnes virtute carentes" (all my worthless ancestors taught me that way too).[31] The heathenish barbarism, the barely cooked flesh eating, and the scatological delight all combine to create the impression of a people so savage as to be beyond redemption.

The effects of the encouragement of the friar and the heathenish barbarity are in Derricke's worldview illustrated in the next wood cut (*Image*, Plate 4, below). Led by their lord, blessed by their friar "if heauen he minde to winne," and previously emboldened by their bard, the Gaelic host sets forth to "plague the princes frendes." But it is in this woodcut that the process of

Figure 10.4 The cattle raid on the settled farmers, Plate 2 from John Derricke's *Image of Irelande* (1581). Special Collections Department, Edinburgh University Library.

Figure 10.5 Mac Sweney feasts, blessed by his friar, Plate 3 from John Derricke's *Image of Irelande* (1581). Special Collections Department, Edinburgh University Library.

the state's retributive violence is introduced. If the Irish are to be brought into line with English civility, a civility that places the queen, "the prince," as the highest authority, and all justice is defined according to the Protestant faith and divine authority, then the "rancoure of the [Gaelic] theeues" must be met with force "of bloudy knife... depriuing them of life."

The implication of woodcuts numbers 4 and 5 (*Image* Plates 4, above, and Plate 5, below) is that combat took place in the open field and only among combatants. Though the savage nature of the "real" combat is indicated by the celebration of decapitation in the accompanying verse: "To see a soldier toze a karne, O Lord it is a wonder! And eke what care he takth to part the head from nexk assonder..."[32] As it is in the visual representation of the same act denote by "A" and "D" in the woodcut.

Though there is clearly a hint in the head-hunting that the warfare practiced in Ireland is not always traditional, woodcuts 6 and 7 (*Image,* Plates 6 and 7, not included here) suggest that Sidney will make a "bloudy fielde" in order to overthrow the "rebels," unless they choose to surrender before his proconsular dignity and powerful military machine represented here in disciplined martial array leaving Dublin Castle. Civil discipline, backed by military might, is visualized in woodcut 7, as Sidney stoops to accept a message from a Gaelic messenger. The savage and heathen Irish will be given one more chance by the queen's representative:

> But, ere they enter mongst those broyles, Syr Henry doth prefarre,
> If happ to get a blessed peace, before most cruell warre,
> Which if they will not take in worth, the folly is their owne,
> For then he goeth with fire and sworde to make her power knowne.

A very specific version of Sidney's sense of himself, the power of the state, and the legitimate martial might of the state is made "real" in woodcut 8 (*Image*, Plate 8, below). In contrast to the earlier representations of anarchic, rebellious, and savage Gaeldom, Sidney's army sets forth in martial array and discipline to "to pronounce by heavy doome, the enemies pryde to lay."

In order to make the civil world real, Derricke visualizes as a prelude a formal battle scene in woodcut 9 (*Image*, Plate 9, below), where it is suggested that the "greasye glibbed foes" will meet their fate at the hand of "Dame Iustice." Yet despite his effort to suggest that Sidney's battle is against the wood kerne, as described in the textual section of the work, woodcut includes in the right foreground elite Gaelic horsemen usually drawn from the ruling family. And despite the fact that there are Gaelic footsoldiers fleeing into the woods in the center background, Derricke may have even included a lined company of Old English horsemen on the enemy side.

Derricke and Sidney's worldview was based on the assumption that they represented civility, justice, and divine retribution. And that as in

A NOTA-
BLE DISCO
uery moſt liue
ly deſcribing the ſtate and
condition of the Wilde men
in Ireland, properly called Woodkarne, with their actions, and exer-
ciſes wherin they are dayly occupied,
alſo the order of their rebellion and
ſucceſſe of the ſame is likewiſe dete-
cted. Which alſo concludeth with
the comming in of *Thirlaugh Leo-
naugh* the great ONEALE of Ireland,
ſubmitting himſelfe to the right ho-
norable Syr *Henry Sydney*, at what
time he was L.Deputy general there
of the ſayd Land, being in *An.1578.*
Nowe publiſhed and ſet forth by
JOHN DERRICK this preſent
yeare of our Lord. 1581. For plea-
ſure and delight of thoſe, whoſe
mindes in laudable exerci-
ſes are vertuouſly
occupied.

Seene and allowed.

¶ At London printed by Iohn Daye
dwelling ouer Alderſgate 1581.

A

The liuely ſhape of Iryſh karne, moſt perfect behold,
Of man, the maſter, and the boy, theſe pictures doe
Wherein is brauely paynted forth, A natu Iriſh grac
Whoſe like in eu'ry poynt to vewe, hath ſome ſtept in
Marke me the karne that gripes the axe, with his mingand,
Then ſhall you ſay a righter knaue, can neuer in the la

1

As for the reſt ſo trimly dreſt, I ſpeake of them no euill,
In ech reſpect, they are detect, (as honeſt as the deuill.)
As honeſt as the Pope himſelfe, in all their outward actions,
And conſtant like the wauering winde, in their Imaginations,
Which may be prou'de in ſundry partes, hereafter that enſue,

Figure 10.6 Blessed by the friar the Gaelic host sets forth, Plate 4 from John Dericke's *Image of Irelande* (1581). Special Collections Department, Edinburgh University Library.

Figure 10.7 A celebration of decapitation, Plate 5 from John Derricke's *Image of Irelan*de (1581). Special Collections Department, Edinburgh University Library.

Figure 10.8 Marching in warlike wise, set out in battle array, Plate 8 from John Derricke's *Image of Irelande* (1581). Special Collections Department, Edinburgh University Library.

Figure 10.9 Dame Justice must proceed against those that have offended, Plate 9 from John Derricke's *Image of Irelande* (1581). Special Collections Department, Edinburgh University Library.

woodcut 10 (not included)—“The Franticke Foes” would be defeated by the disciplined forces of the state and that Sidney would be greeted by universal joy. Because despite the effort to make visually real a victory, the reality was much more bitter for the Sidney camp. In the accompanying verse Derricke suggests: “When thus this thrice renowmed Knight, hath captiue made and thrall, The furious force of franticke foes, and troupe of rebells all . . . ,” he returned to Dublin, where he was “receiued with ioy on euery parte.” This was highly unlikely given the fact that if it was in 1578 before he was recalled in disgrace it was to a Pale sullen with resentment over the enormous costs involved in his military campaigns and the dispute over the cess, the taxation system imposed to pay for it, and before the actual death of Rory Óg O’More.

In contrast to the realities of Sidney’s recall in disgrace in March 1578, Derrick attempted to build a world where retributive violence worked in the hands of his and its patron, Sir Henry Sidney. This is why in woodcut 11 (*Image,* Plate 11, below), the second last in the series, he visualizes the defeated rebel Rory Óg O’More hunted and despised, alone in the woods about to meet his end at the hands of the forces of law and order. His challenge to the queen’s authority, his sin, in Derricke’s terms “he sinde in that he moued our noble Queene to ire,” would inevitably result in his miserable demise. He is depicted in the woodcut as miserable and in the woods. Alone and wrapped in his mantle he states “ve mihi misero” (woe to wretched me) while his vicious companion the wolf responds in agreement, “ve atque dolor” (woe and grief), both foretelling his miserable end “tombled in the myre.”[33]

Derricke and by extension Sidney justified state sponsored violence on behalf of the queen’s authority, and the words that he puts into Rory Óg’s mouth suggest that they wanted to demonstrate that the heathen Irish, if only at the point of death, had seen and understood the English worldview to be right. In much the same way as the final woodcut (*Image,* Plate 12, below), number 12, showed Ireland’s leading Gaelic lord, Turlough Luineach O’Neill, the head of the great O’Neill lordship of Ulster, bowed down before a sitting and somewhat regal Sidney, with the sword of state to his right on a cushion. The sword here also represents the sword of Justice and of the retributive power of the prince Sidney came to uphold: “in Inglands claime of Iustice there he came; And to maintayne the sacred right of such a Virgine Quenne . . .”

Yet the woodcut of the mighty O’Neill’s submission also seems to inordinately celebrate Sidney, though acknowledging that the deputy stands in for the prince in verse: “Of good Syr Henry Sydney . . . Loe where he sittes in honours seate, moste comely to be seene, As worthy is to represent the person of a Queene.” This particular image of a resplendent Sidney in an elaborate and ceremonial tent seems to this reader to hint at the notion that the Sidneys, father and son, may have felt that they were “as worthy,” given their enormous effort, as the prince herself.

The woodcut of O'Neill's submission visually represented the cumulative effectiveness of force, of coercion, wielded by a master statesman. Force could crush the heathen "Catholic" rebel, and the witnessing of brutal effectiveness could bring the recalcitrant to kneel before Sidney and the symbols of the prince and the civil state. Thus in Derricke's images and text, and by extension Sidney's worldview, as represented in both sections of the *Image of Irelande,* the "reform" of Ireland happens as a result of coercion, not as an alternative to it.[34]

Derricke's woodcuts clearly celebrate violence but also in a way they sanitize it. There is ultimately a disconnect between the verse section on the campaign against O'More and the images, suggesting the possibility that the former was originally drawn up before the vehement campaigning of the early months of 1578. As the war against Rory Óg O'More unfolded, Derricke added to a text composed originally in Parts I and II as a poetic discourse on Gaelic society, typed as wood kern. The text is transformed in the fictive account of Rory's life and rebellion (sigs. Hii–Liii) into a celebration of a victory over a notable rebel. Augmented yet again with an update on Rory's death, we get a sense of an author, a Sidney client, eagerly awaiting more news of his patron's successes in order to trumpet his success in print. That the work was not published in whole until 1581 is a problem yet to be solved, but that it coincided with a Sidney campaign to have his government service vindicated and possibly for a reappointment to the Irish lord deputyship should not be discounted. Published at the height of the Desmond rebellion, Sidney can only have been eager to show that he could bring major rebels to boot. What galled him the most about his success in this regard was that his crushing of Rory Óg O'More paid for with a high price in lives was denigrated. For the queen made "so little of my service in killing that pernicious rebel and was contented to be persuaded that there was no more difficulty to kill such a rogue as he was to kill mad George the sweeper of the queen's court . . ."[35] Sidney wanted the political world to know and understand both in his "Memoir" and by extension in Derricke's *Image* that Rory Óg was a dangerous monster, a monster that he had decapitated. What separated the two works was an emphasis in the latter on the barbarity of the Irish and also a celebration of violence. One was semiprivate, the "Memoir," the other was public, and it in the latter the Sidney camp wanted the world to know that he could hunt with the best of them. Hence the inclusion of the material on Rory Óg, and the separate dating of the woodcut section of the work, for here in 1581, as rumors of Arthur Grey's recall whirled around court; the appended visual apologia attempted to suggest that the Sidney house—Philip may have been a candidate too—understood the Irish and knew best how to chastise them.

The woodcuts that accompanied Derricke's *Image of Irelande* operated as a visual resume advertising the effectiveness of Sidney as a governor and celebrated retributive violence. That the images coincide with the text to endorse violence against the Gaelic Irish should not be doubted. Suggestions that these work together to support a Sidney "program" of reform should be qualified. As Derricke awaited news from the Midlands he gave vent to an expression of frustration with his literary project but also with dealing with a people like the O'Mores, and hoped for the "finall ending of this wretched race." Sidney his patron did his best to accomplish this task. Boasting of a headcount of over 750 of these "rebels," Sidney laid such a hand on the leading *Sliocht* or sept of the Mac Conaill O'Mores that by the time he left, vast majority of the leading males had met a violent death through hanging, decapitation, or in combat. If atrocity was a necessary tool in or prelude to the "reform" of the Gaelic Irish, then let us recognize it as such. Sidney despite claims to the contrary was not adverse to benefits of unrestrained violence and even ethnic cleansing in order to advance English rule in Ireland. Derricke's *Image of Irelande* recognizes this worldview and indeed celebrates it.[36]

Notes

1. Sir Henry Sidney, *A Viceroy's Vindication? Sir Henry Sidney's Memoir of Service in Ireland, 1556–78*, ed. Ciaran Brady (Cork: Cork University Press, 2002), 109; hereafter referred to as "Memoir."
2. The term "retributive justice" is Brady's, see, "Memoir," 101, and n. 118.
3. Vincent P. Carey, "John Derricke's *Image of Irelande*, Sir Henry Sidney, and the massacre at Mullaghmast, 1578," *Irish Historical Studies* 31 (1999): 305–327.
4. "Memoir," 35.
5. I am extremely grateful to Professor Maryclaire Moroney for a critical reading of my draft and for helping clarify my thinking here.
6. Maryclaire Moroney, "Apocalypse, Ethnography and Empire in John Derricke's *Image of Irelande* (1581) and Spenser's *View of the Present State of Ireland* (1596)," *English Literary Renaissance* 29, no. 3 (1999): 355–374; Carey, "John Derricke's *Image of Irelande.*"
7. Edmund Spenser, *A View of the Present State of Ireland*, ed. Andrew Hadfield and Willy Maley (Oxford: Blackwell, 1997).
8. "Memoir," 34.
9. Willy Maley, " 'The Name of the Country I Have Forgotten': Remembering and Dismembering in Sir Henry's Irish *Memoir* (1583)," in Thomas Herron and Michael Potterton, eds. *Ireland in the Renaissance c. 1540–1660* (Dublin: Four Courts, 2007), 52–73.
10. James A. Knapp, *Illustrating the Past in Early Modern England: The Representation of History in Printed Books* (New York: Ashgate, 2003).
11. Ibid., 207.
12. Ibid., 210.
13. Ibid., xxx.
14. Ibid., 212.

15. Ibid., 212.
16. Ibid., 220–228.
17. Ibid., 229.
18. Richard McCabe, *Spenser's Monstrous Regiment: Elizabethan Ireland and the Practice of Difference* (Oxford: Oxford University Press, 2002), 74.
19. *Image*, sig. aii.
20. Ibid., sig. aiiv.
21. Ibid., sig. aii–aiiv.
22. "Sidney to the Lords of the Council in England," July 1, 1578, in Arthur Collins, *Letters and Memorials of State*, vol. 1, 263–264.
23. A sense of accomplishment that Sidney in bitterness in the memoir claims was diminished by his detractors at court.
24. *Image*, sig. Hi.
25. Ibid., sig. Hiv.
26. McCabe, *Spenser's Monstrous Regiment*, 74.
27. The image of Sidney taking the submission of Turlough Luineach O'Neill.
28. It should be noted that the saddle is a cushion and there are no stirrups; see the detailed notes of D.B. Quinn to his edition of Derricke's *Image*, where he thoroughly explores and contextualizes the various images included in the woodcuts; D.B. Quinn, ed. *The Image of Irelande with a Discoverie of Woodkarne by John Derricke 1581* (Belfast: Blackstaff Press, 1985).
29. Louis Montrose, "The Work of Gender in the Discourse of Discovery," in Stephen Greenblatt, ed. *New World Encounters* (Berkeley: University of California Press, 1993), 177–217.
30. Alan J. Fletcher, *Drama, Performance and Polity in Pre-Cromwellian Ireland* (Toronto: University of Toronto Press, 2000), 16–17. I am grateful to Thomas Herron for this reference.
31. Translation is by D.B. Quinn in Notes to Plate 3.
32. On beheading, see the excellent article of Patricia Palmer, " 'An Headless Ladie' and 'a horses loade of heades': Writing the Beheading," *Renaissance Quarterly* 60 (2007): 32–34.
33. John Barry, "Derrick and Stanihurst: A Dialogue," in Jason Harris and Keith Sidwell, eds. *Making Ireland Roman: Irish neo-Latin Writers and the Republic of Letters* (Cork: Cork University Press, 2009), 36–27 and 209, n. 30.
34. I am grateful to Maryclaire Moroney's clarification of my thinking here.
35. "Memoir," 102.
36. I would like to thank Clare Carroll, Sarah Covington, Valerie McGowan-Doyle, Thomas Herron, Brendan Kane, Hiram Morgan, and Maryclaire Moroney for assistance with this chapter. Needless to say the views expressed are my own.

Contributors

Vincent Carey is Professor of History at SUNY Plattsburgh. His works focus on the violence inherent in the English conquest of Ireland. His recent books include *Surviving the Tudors: Gerald the "Wizard" Earl of Kildare and English Rule in Ireland, 1537–1586* (Four Courts Press, 2002) and a coedited volume with Clare Carroll of *Solon, His Follie* (Medieval Renaissance Texts and Studies, 1996). He was the creator of the Folger Shakespeare Library exhibition on early modern atrocities in 2002 and 2003. He is currently working on a book about atrocities and atrocity literature in early modern Ireland.

Al Coppola is Assistant Professor of English at John Jay College of the City University of New York. His current book project, *The Theater of Experiment: Staging Natural Philosophy in Eighteenth-Century Britain,* explores the role of spectacle in the production of natural facts, as well as the ways in which science was itself performed in both domestic and theatrical spaces. His study of the science and politics of Aphra Behn's *Emperor of the Moon* appeared in *Eighteenth-Century Studies,* and an article tracing the links between the *Harlequin Doctor Faustus* pantomimes and Newtonian public science lectures in the 1720s appears in a University of Delaware Press volume devoted to the eighteenth-century actor and theatrical manager John Rich.

Dane T. Daniel is Associate Professor of History at Wright State University, Lake Campus. He has been awarded fellowships at the Dibner Institute for the History of Science and Technology, and the Chemical Heritage Foundation. He won the 2005 Partington Prize, awarded every three years by the Society for the History of Alchemy and Chemistry (London), and has published a number of articles and book chapters, including "Medieval Alchemy and Paracelsus' Theology: Pseudo-Lull's *Testamentum* and Paracelsus' *Astronomia Magna,*" *Nova Acta Paracelsica* (2009): 121–136; "Invisible Wombs: Rethinking Paracelsus's Concept of Body and Matter," *Ambix* (2006): 129–142; and "Paracelsus on the Lord's Supper: *Coena Dominj Nostrj Jhesu Christj Declaratio.* A Transcription of the Leiden Codex Voss. Chym. Fol. 24, f. 12r–29v." *Nova Acta Paracelsica* (2002): 107–139.

James De Lorenzi is Assistant Professor of History at CUNY John Jay College of Criminal Justice. He is a broadly trained historian of empire with a focus on the modern Red Sea region, and his work has been published in the *Journal of World History* and *Comparative Studies of South Asia, Africa, and the Middle East*. He is currently serving as Assistant Editor of the journal *Northeast African Studies*.

Guido Giglioni is Cassamarca Lecturer in Neo-Latin Cultural and Intellectual History at the Warburg Institute, University of London. He has published a book on Jan Baptise van Helmont (*Immaginazione e malattia*, Milan, 2000) and edited a volume of manuscript papers of Francis Glisson (1996). He has written essays on Renaissance philosophy and medicine and on such authors as Girolamo Cardano, Tommaso Campanella, and Francis Bacon.

Allison B. Kavey is Associate Professor of History at CUNY John Jay College of Criminal Justice and the CUNY Graduate Center. She is the author of *Books of Secrets: Natural Philosophy in England, 1550–1600* (University of Illinois Press, 2007) and coeditor with Lester D. Friedman of *Second Star to the Right: Peter Pan in the Cultural Imagination* (Rutgers University Press, 2008). She is currently working on a book about the magical cosmology in Agrippa von Nettesheim's *De Occulta Philosophia Libri Tres*.

Sheila J. Rabin is Professor of History at Saint Peter's College in Jersey City, New Jersey. She writes on the Renaissance debate on astrology, focusing on the works of Giovanni Pico della Mirandola and Johannes Kepler. Among her publications are "Kepler's Attitude toward Pico and the Anti-astrology Polemic," *Renaissance Quarterly* 50 (1997): 750–770 and "Was Kepler's *Species Immateriata* Substantial?" *Journal for the History of Astronomy* 36 (2005): 49–56.

Matteo Salvadore received his doctorate in history from Temple University with a dissertation on the history of Ethiopian-European cultural relations in the early-modern and modern era. Among his publications are "A Modern African Intellectual: Gäbre-Heywät Baykädañ's Quest for Ethiopia's Sovereign Modernity" (*Africa*, Roma 2008); "The Ethiopian Age of Exploration: Prester John's Discovery of Europe, 1306–1458," *Journal of World History* 21, no. 4 (December 2010), and a book chapter on the social consequences of Italian colonialism on the colonized to be published in Italy later this year.

Patrick Tuite is an Associate Professor and the Head of the M.A. Program in Theatre History and Criticism in the Department of Drama at the Catholic University of America. He has published articles in *Youth Theatre Journal, Theatre InSight,* and *The Drama Review*. His essays also appear in anthologies entitled *Audience Participation: Essays on Inclusion in Performance* (Greenwood Press, 2003) and *Second Star to the Right: Peter Pan in the Popular Imagination* (Rutgers University Press, 2008). Fellowships at the Folger Shakespeare Library and the Keough-Naughton Institute for Irish Studies at the University of Notre Dame supported Dr. Tuite's research for this essay and his book, *Theatre of Crisis: The Performance of Power in the Kingdom of Ireland, 1662–1692* (Susquehanna University Press, 2010).

Mark A. Waddell is a Visiting Assistant Professor in the Lyman Briggs College at Michigan State University. He has written and published extensively on the role of the Jesuits in early modern intellectual culture, with a particular focus on Athanasius Kircher. He is currently at work on his first book, *Light and Shadow: The Jesuit Study of the Invisible and the Crisis of Uncertainty*.

Index